About Island Press

Since 1984, the nonprofit organization Island Press has been stimulating, shaping, and communicating ideas that are essential for solving environmental problems worldwide. With more than 1,000 titles in print and some 30 new releases each year, we are the nation's leading publisher on environmental issues. We identify innovative thinkers and emerging trends in the environmental field. We work with world-renowned experts and authors to develop cross-disciplinary solutions to environmental challenges.

Island Press designs and executes educational campaigns in conjunction with our authors to communicate their critical messages in print, in person, and online using the latest technologies, innovative programs, and the media. Our goal is to reach targeted audiences—scientists, policymakers, environmental advocates, urban planners, the media, and concerned citizens—with information that can be used to create the framework for long-term ecological health and human well-being.

Island Press gratefully acknowledges major support from The Bobolink Foundation, Caldera Foundation, The Curtis and Edith Munson Foundation, The Forrest C. and Frances H. Lattner Foundation, The JPB Foundation, The Kresge Foundation, The Summit Charitable Foundation, Inc., and many other generous organizations and individuals.

Generous support for the publication of this book was provided by Terry Gamble Boyer and Peter Boyer.

The opinions expressed in this book are those of the author(s) and do not necessarily reflect the views of our supporters.

Designing Climate Solutions

Designing Climate Solutions:
A Policy Guide for
Low-Carbon Energy

By Hal Harvey

With Robbie Orvis, Jeffrey Rissman, Michael O'Boyle,
Chris Busch, and Sonia Aggarwal

ISLANDPRESS

Washington | Covelo | London

Library of Congress Control Number: 2018946759

All Island Press books are printed on environmentally responsible materials.

Manufactured in the United States of America

10 9 8 7 6 5 4 3

Keywords: appliance standards, building codes, carbon emissions, carbon pricing, carbon tax, climate change, decarbonization, economic signals, electric grid, electric vehicles, energy efficiency, energy-efficient buildings, energy policy, energy technology, energy use, feed-in tariffs, greenhouse gases, industry, Paris Accord, performance standards, energy, renewable portfolio standards, renewable wind power, research and development, solar power, transportation policy, urban design, vehicle efficiency standards

Contents

Foreword

The 1992 Rio de Janeiro declaration embodied international recognition that global warming and climate change, driven by anthropogenic greenhouse gas (GHG) emissions, pose a significant risk to the world's economy, environment, and security and demand a cooperative effort between nations. It is often forgotten in today's debate in the United States that the Senate ratified this agreement and thereby committed our country to address this challenge. Nearly a quarter century later, at the historic Paris Conference of the Parties in late 2015, numbers were attached to the broad Rio declaration: Nations, both developed and developing alike, put forward specific goals for GHG emission reductions in the 2025–2030 timeframe. These national goals reflect the enormous scientific understanding developed in recent decades. With widespread compliance, they represent a reasonable path toward meeting an overarching goal: limiting global warming to two degrees Celsius, or possibly less. The two-degree goal would call for meeting much more ambitious GHG emission reduction goals beyond 2030. Presumably the industrialized countries, with large per capita GHG emissions, would be called on to make especially significant early reductions of their energy sector emissions.

In fact, progress has been made. In the United States, both market forces (through the substitution of natural gas—and increasingly renewables—for coal) and federal and state clean energy requirements have substantially reduced carbon dioxide emissions. China has leveled its coal use. Many other countries have introduced and implemented policies to reduce emissions, including by the imposition of carbon emission charges, although much more needs to be done across the board. Furthermore, the June 2017 announcement by President Trump that he will pull the United States out of the Paris agreement sowed confusion as to where the world's second-largest emitter (after China) was headed. Fortunately, state governors, city mayors, and business leaders soon made clear that they fully anticipated continued progress toward a low-carbon future and would stay the course or even accelerate their plans. We are not going back.

The issue is how we get there and how fast. Because of the cumulative nature of carbon dioxide emissions, time is of the essence. A failure to act decisively now exacerbates the challenge in the decades ahead. Success will require synergistic and innovation in technology, business models, policies, and regulations.

Hal Harvey and his coauthors have performed an important service in *Designing Climate Solutions: A Policy Guide for Low-Carbon Energy*. In this work, the focus is placed on *how* to reach these emission targets and which policies have a reasonable chance of getting us there. They have relied on decades of experience combined with quantitative analysis to present a portfolio of policy solutions for key climate change risk mitigation opportunities that have been shown to work in a variety of contexts and countries. The chapters thus serve as a handbook for policymakers on both national and subnational levels. *Designing Climate Solutions* addresses policies that provide economic signals, performance standards, and R&D support. Clearly, this suite of policy models needs adaptation to local, regional, and national circumstances, just as there is no single low-emission technology solution for different localities and countries. This book provides a valuable toolkit for policymakers committed to timely mitigation of climate change risks.

Harvey and colleagues promote pragmatic optimism for reaching the challenging two degree Celsius goal. In their approach, they embody the philosophy of Bostonian Willie Sutton, who is said to have answered the question of why he robbed banks with the response, "Because that's where the money is." Harvey and his coauthors emphasize that only 7 countries are responsible for more than half of the GHG emissions, and only 20 for three-quarters. So "that's where the carbon is." Thus, a small portfolio of proven policies, applied in a small number of countries, can yield enormous progress toward the global challenge of limiting global warming and climate change. The imperative is to move expeditiously, and this book provides an excellent foundation for effective policy design in diverse circumstances.

Ernest J. Moniz
13th U.S. Secretary of Energy

Acknowledgments

The authors thank the following for their help in writing, reviewing, and editing this book:

Don Anair, Galen Barbose, Kornelis Blok, Dale Bryk, Dallas Burtraw, Rachel Cleetus, Christine Egan, Seth Feaster, Jamie Fine, Ben Friedman, John German, Justin Gillis, Eric Gimon, Bill Hare, Devin Hartman, Sara Hastings-Simon, C. C. Huang, Hallie Kennan, Drew Kodjak, Charles Komanoff, Honyou Lu, Silvio Marcacci, Cliff Maserjik, Matt Miller, Erica Moorehouse, Dick Morgenstern, Simon Mui, Colin Murphy, Steven Nadel, Stephen Pantano, He Ping, Conor Riffle, Richard Sedano, Jigar Shah, Jessica Shipley, Kelly Sims-Gallagher, Robert Sisson, Heather Thompson, Zachary Tofias, Michael Wang, and Fang Zhang.

Introduction

To put the world on a path to a reasonable climate future, immediate action is needed to reduce greenhouse gas emissions. The mounting evidence of potential damage from climate change is daunting, and with each day that passes the challenge ahead becomes more difficult. At the same time, new technologies continue to show that a low-carbon future is within reach and perhaps as cheap as or cheaper than a high-carbon one.

Reducing global greenhouse gas emissions is no small task. But the technologies, policies, and motivation to achieve this reduction exist today; it is a matter of adopting, designing well, and then promptly implementing the right policies.

The vast majority of greenhouse gas emissions come from a small set of countries; their source is predominantly energy use, such as power plants, vehicles, and buildings, and industrial processes, such as cement or iron and steel manufacturing. Focusing on energy use and industrial processes has the largest potential for emission abatement.

Fortunately, a small set of policies exist that have the potential to significantly reduce emissions from these sectors. For example, vehicle efficiency standards, which require vehicle manufacturers to increase the distance vehicles can travel on the same amount of fuel, can rapidly drive down emissions from transportation, and policies to promote the share of carbon-free electricity, such as renewable portfolio standards and feed-in tariffs, can reduce emissions in the power sector. A dozen highly effective policies in the biggest countries can put us on the right path.

Of course, these policies must be designed well if they are to achieve lasting reductions. Decades of experience with both good and bad policy design has illuminated the characteristics that separate good from bad policy. For example, without built-in mechanisms for continuous improvement, policies tend to stagnate and become obsolete. And without a sufficiently long time horizon, businesses cannot invest in the technology or research and development (R&D) needed to produce better equipment. A handful of policy design principles can ensure that future climate and energy policy maximizes greenhouse gas reductions and economic efficiency. These policies leverage the trillions of dollars of private capital spent each year, already, to build a clean energy policy. In other words, these principles can drive effective, investment-grade policy.

This book drills down into these policies, their design principles, and their potential impact on global emissions. Our hope is that this material can serve as a resource to policymakers, CEOs, nongovernment organizations, research institutions, and philanthropists who are searching for the fastest, most effective way to make a big difference in reducing the threat of climate change.

In Climate, Delay Is Killer

There is broad consensus that preventing the worst impacts of climate change requires keeping global warming below two degrees Celsius through the end of the 21st century. This, in turn, requires steep cuts in greenhouse gas emissions. Climate models vary, but to have an even chance of staying under two degrees, we need to avoid 25 to 55 percent of cumulative emissions between now and 2050 compared with the business-as-usual case.[1] The needed reductions vary significantly by region, with steeper reductions needed for more advanced economies. Furthermore, whereas cumulative emission reductions of 25 to 55 percent are required, annual emissions in 2050 must be lower still, on the order of 40 to 70 percent below business-as-usual emissions (Figure I-1).

The scope, scale, and irreversibility of climate change—and the irreducible mathematics of carbon accumulation—together mean that swift action to abate greenhouse gas emissions is imperative. Failing to take immediate action to reduce emissions could result in significant damage: loss of coastal lands to sea level rise, threatening more than a billion people; mass refugee migration; famines; a wave of extinctions; and other impacts that will take an economic, ecological, and human toll. Rising seas in Bangladesh, for example, could produce 35 million refugees,[2] seven times the number generated by the crisis in Syria, which has shaken the political stability of Europe. Scientists predict up to a 40 percent reduction in East Africa's wheat and maize production from heat alone.[3] It is a grim list—and a long one.

The seriousness of the threat and the limited time to tackle it are both a consequence of scientific facts about Earth's biogeochemical systems. Small shifts in average global temperature have outsize consequences. Three important factors drive this magnification: the way in which those shifts increase the frequency of extreme temperature and weather events, the irreversibility of warming on reasonable timescales, and the danger of triggering natural feedback loops that cause additional warming.

Extremes Become the Norm

Any given place on Earth experiences a range of temperatures, both day to day and year to year. Consider the average summer temperature in the United States: In certain years, the country experiences unusually cool summers, and in other years it experiences particularly hot summers. But in most years, the summer temperature is around average for that time of year.

Increasing the *average* global temperature makes previously rare *extreme* temperatures become much more frequent. This has the effect of making cooler summers rare and really hot summers more common (Figure I-2).

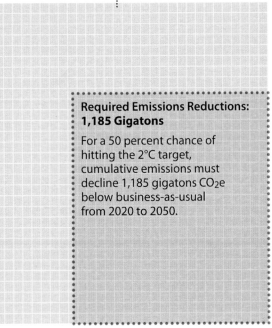

Business As Usual Cumulative Emissions, 2020-2050: 2,253 Gigatons CO$_2$e

Required Emissions Reductions: 1,185 Gigatons

For a 50 percent chance of hitting the 2°C target, cumulative emissions must decline 1,185 gigatons CO$_2$e below business-as-usual from 2020 to 2050.

Figure I-1. Emission reductions needed for a 50 percent chance of avoiding 2°C global warming. (Analysis done using data with permission from the International Institute for Applied Systems Analysis [IIASA]. Data source: M. Tavoni, E. Kriegler, T. Aboumahboub, K. Calvin, G. De Maere, J. Jewell, T. Kober, P. Lucas, G. Luderer, D. McCollum, G. Marangoni, K. Riahi, and D. van Vuuren, "The Distribution of the Major Economies' Effort in the Durban Platform Scenarios," *Climate Change Economics* 4, no. 4 (2013), doi:10.1142/S2010007813400095. Data downloaded from the LIMITS Scenario database hosted at IIASA, https://tntcat.iiasa.ac.at/LIMITSPUBLICDB/dsd?Action=htmlpage&page=about.)

We are already starting to experience these effects, as exceptionally hot summers that occurred less than once every three hundred years or so from 1951 to 1980 represented a significant fraction of all summers from 2005 to 2015 (Figure I-3).

Temperature: Increase in mean

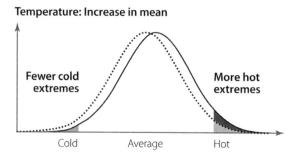

Temperature: Increase in variance

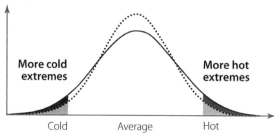

Temperature: Increase in mean and variance

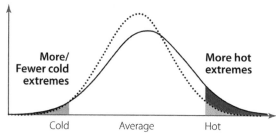

Figure I-2. Higher and more variable temperatures lead to greater temperature extremes. (Graphic reproduced with permission from the Intergovernmental Panel on Climate Change. Data from Figure 1.8 from Cubasch, U., D. Wuebbles, D. Chen, M.C. Facchini, D. Frame, N. Mahowald, and J.-G. Winther, 2013: Introduction. In *Climate Change 2013: The Physical Science Basis. Contribution of Working Group I to the Fifth Assessment Report of the Intergovernmental Panel on Climate Change* [Stocker, T.F., D. Qin, G.-K. Plattner, M. Tignor, S.K. Allen, J. Boschung, A. Nauels, Y. Xia, V. Bex, and P.M. Midgley (eds.)]. Cambridge University Press, Cambridge, United Kingdom and New York, NY, USA.)

This effect is projected to become much worse in the future. Even with an average global increase of just a few degrees, many areas in the United States that today see just a few days with highs of at least 100°F will see many more such days by the end of the century. For example, "By the middle of this century, the average American will likely see 27 to 50 days over 95°F each year—two to more than three times the average annual number of 95°F days we've seen over the past 30 years. By the end of this century, this number will likely reach 45 to 96 days over 95°F each year on average."[4] In other words, by the end of this century, sections of Texas, Arizona, and California could swelter in such conditions for a third of the year or more (Figure I-4).

These exceptionally hot summers cause severe damage, because they are beyond the typical range to which human and natural systems have adapted. For example, exceptionally hot conditions dry out the landscape, intensify wildfires, devastate crop and livestock yields, send people to the hospital with heat stroke, and cause many other harms.

The Irreversibility of Warming on Reasonable Timescales

Once a quantity of greenhouse gas is emitted, it will begin cycling out of

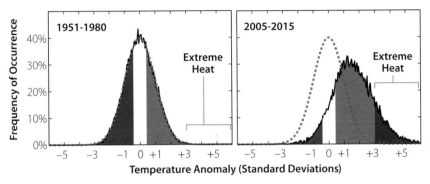

Figure I-3. Climate change is shifting global temperatures, making extreme heat more frequent. (Reproduced from publicly available U.S. government data, from James Hansen et al., "Public Perception of Climate Change and the New Climate Dice," n.d., http://www.columbia .edu/~mhs119/PerceptionsAndDice/.)

the system as various natural cycles pull it out of the atmosphere. For example, carbon dioxide (CO_2), the most important greenhouse gas, may dissolve into the ocean, and methane eventually breaks down into CO_2. For many greenhouse gases, it takes a very long time to remove nearly all the pollutant from the atmosphere. The natural removal rate for CO_2 is slow—it may take hundreds or thousands of years without human emissions for CO_2 to return to its natural concentration. Likewise, it may take hundreds of years to eliminate almost all atmospheric nitrous oxide (a strong greenhouse gas) and many thousands of years for various fluorinated gases, which are very strong greenhouse gases. Indeed, much of the carbon dioxide emitted at the very start of the Industrial Revolution—about 250 years ago—is still present in the atmosphere today.

The climate system also has a great deal of inertia. That is, climate change is a problem of stocks, not flows. One way to visualize this is to think of the atmosphere as a bathtub (Figure I-5): As carbon dioxide is emitted (the faucet), it continues to add to the total carbon dioxide concentration in the atmosphere (the bath water). Carbon dioxide is then removed by natural processes over many years (the drain). Like a bathtub in which the water level will continue to rise as long as there's more water flowing in than draining out, even a dramatic reduction in carbon dioxide *emissions* will not reduce *concentrations* as long as emissions outpace removals.[5]

Even if we were to completely stop emitting greenhouse gases today, the impacts of the previously emitted gases would continue to be felt for thousands

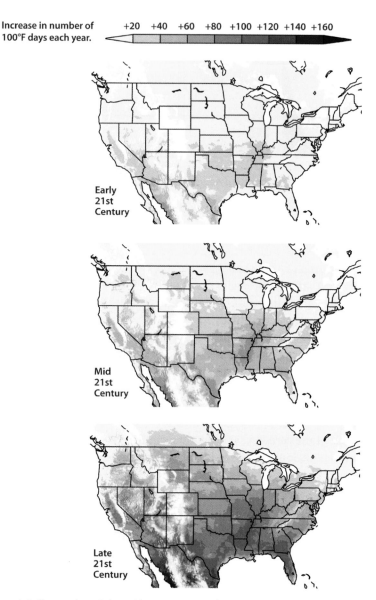

Figure I-4. The number of days with temperatures above 100°F may increase significantly as climate change worsens. (Reproduced from publicly available U.S. government data, from "LOCA Viewer," Scenarios for the National Climate Assessment, accessed January 8, 2018, https://scenarios.globalchange.gov/loca-viewer/.)

of years, because the total stock of carbon in the atmosphere (the bath water) remains high. Once in the atmosphere, CO_2 and other greenhouse gases trap heat in the atmosphere, and the effects of that trapped heat manifest over time. As a result, we will continue to see increasing impacts on human society and natural systems for thousands of years after stabilizing the greenhouse gas concentration in the atmosphere.

Note in Figure I-6 that greenhouse gas *emissions* (in carbon dioxide equivalent, or CO_2e), which are caused by human activity, are driven to near zero, while greenhouse gas *concentrations* barely decline, and greenhouse gas *impacts* (temperature) keep growing.

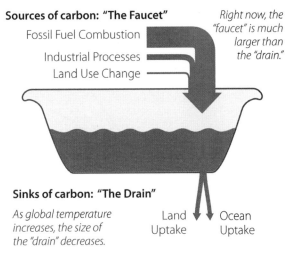

Sources of carbon: "The Faucet"

Fossil Fuel Combustion
Industrial Processes
Land Use Change

Right now, the "faucet" is much larger than the "drain."

Sinks of carbon: "The Drain"

As global temperature increases, the size of the "drain" decreases.

Land Uptake   Ocean Uptake

Figure I-5. Greenhouse gas emissions and concentrations are like a bathtub, with emissions being the faucet, concentration being the tub, and sinks being the drain. (From "Causes of Climate Change," U.S. EPA, 2016, https://19january2017snapshot .epa.gov/climate-change-science/causes-climate-change_.html.)

The Danger of Natural Feedback Loops

One of the most disturbing aspects of the global warming problem is that as the world heats up, natural feedback loops kick in that intensify the warming. Although anthropogenic (human-caused) emissions may be the initial catalyst in warming the globe, Earth's natural systems can exacerbate this impact, creating what physicists call a positive feedback loop, which is more easily understood as a vicious cycle. One vicious cycle is the impact melting sea ice has on Earth's absorption of heat: Bright sea ice has a high albedo, meaning it reflects (rather than absorbs) most of the light that hits it. Dark sea water has a lower albedo, meaning it absorbs the light that hits it, which turns into heat. As ice melts from warming temperatures, areas previously covered with reflective white ice become uncovered with absorptive blue ocean water, which increases heat absorption and further accelerates warming.

A similar vicious problem occurs with melting arctic tundra: As once-frozen tundra thaws from warmer temperatures, buried methane deposits are released, causing more greenhouse gases to enter the atmosphere and warm

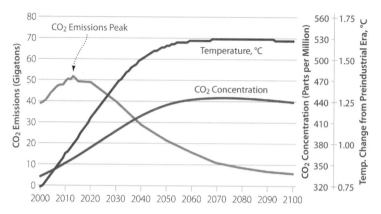

Figure I-6. Even if emissions were to peak and drop to zero immediately, CO_2 concentrations and temperatures would continue to increase. (Reproduced with permission from the International Institute for Applied Systems Analysis [IIASA]. Data source: Clarke L., K. Jiang, K. Akimoto, M. Babiker, G. Blanford, K. Fisher-Vanden, J.-C. Hourcade, V. Krey, E. Kriegler, A. Löschel, D. McCollum, S. Paltsev, S. Rose, P.R. Shukla, M. Tavoni, B.C.C. van der Zwaan, and D.P. van Vuuren, 2014: "Assessing Transformation Pathways." In: *Climate Change 2014: Mitigation of Climate Change. Contribution of Working Group III to the Fifth Assessment Report of the Intergovernmental Panel on Climate Change* [Edenhofer, O., R. Pichs-Madruga, Y. Sokona, E. Farahani, S. Kadner, K. Seyboth, A. Adler, I. Baum, S. Brunner, P. Eickemeier, B. Kriemann, J. Savolainen, S. Schlömer, C. von Stechow, T. Zwickel, and J.C. Minx (eds.)]. Cambridge University Press, Cambridge, United Kingdom and New York, NY, USA. Data downloaded from the IPCC-IAMC database hosted at IIASA, https://secure.iiasa.ac.at/web-apps/ene/AR5DB.)

the world further. The scale of this greenhouse accelerator, once released, is almost unfathomable, and once it starts, it cannot be controlled.

Yet another is the absorption of CO_2 in oceans: As oceans absorb more CO_2, water becomes more acidic, causing the die-off of aquatic plants and animals, whose decomposition contributes additional CO_2 to the oceans and atmosphere.

It is unclear exactly how much these feedback loops will exacerbate climate change. But their potential to accelerate warming is frightening—and the fact that these forces become uncontrollable once unleashed means climate action is necessary immediately.

Delay Is Costly

Another reason for acting as soon as possible to reduce emissions is that the challenge of cutting emissions enough to avoid exceeding two degrees of

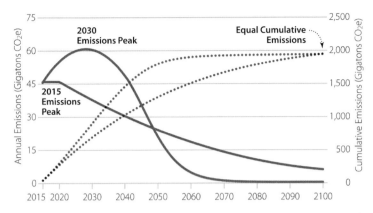

Figure I-7. The longer the delay in peaking emissions, the harder it becomes to meet the same carbon budget.

warming will get increasingly difficult as reductions are delayed. Delaying is costly for two reasons. First, most energy-consuming assets—buildings, power plants, industrial facilities—have a turnover rate of decades or more, meaning we essentially lock in a higher level of warming with each piece of new equipment we adopt or install. Second, because warming is a function of the total amount of carbon dioxide in the atmosphere, delayed action on emission reductions makes it far harder to achieve the same concentration of CO_2 in the future.

For example, Figure I-7 shows how a 15-year delay in peak emissions requires greater emissions reductions, faster, to achieve the same cumulative level of emissions. Note that not only is the emission reduction rate higher in each subsequent year, but the total level of emissions, particularly in the latter half of the century, must be lower to account for the higher emissions early on. The sooner action is taken, the easier it will be to meet the two-degree target. Waiting even a few years can significantly exacerbate the challenge of keeping warming below two degrees.

Delay comes with significant costs, too. An analysis by eminent British economist Lord Nicholas Stern in the *Stern Review* shows that stabilizing greenhouse gas concentrations at 500–550 parts per million (about 100–150 parts per million higher than today's concentration and nearly double the pre–Industrial Revolution concentration of about 280 parts per million) would cost approximately 1 percent of global economic output (gross domestic product

[GDP]) per year, whereas the costs of inaction would mount to 5–20 percent of global GDP per year from the costs of climate change impacts.[6]

The physics of our Earth thus give us the following imperatives: The problem is enormous, it is urgent, and failure would be irreversible. Fortunately, there is still time to achieve a reasonable climate future and many reasons to think it can be done. But time is of the essence; this option does not last long.

Reasons for Hope

The effects of climate change are worrisome, to say the least. Emission reductions are needed as soon as possible to avoid the worst of these effects. Fortunately, there is ample technology to put the world on a low-carbon trajectory. It is backed by growing political momentum.

Cleantech: From Boutique to Mainstream

Renewable Energy

Transitioning to a low- or zero-carbon electricity system is no longer a dream of the future. Costs for wind and solar power have plunged, propelling their growth around the world. Contracts for U.S. wind projects have been coming in at less than half the price they were just 5 years ago and cheaper than any other new power source worldwide. Contracts for solar power in some parts of the world are coming in at the same or a better price.

To put these cost declines into context, consider that recent projects in Chile and Dubai have been contracted at less than 3 cents per kilowatt-hour, without any subsidy, compared with residential electricity rates in the United States of nearly four times that amount. The costs of solar photovoltaic systems are projected to fall even further—below $1.00 per watt by 2020[7]—and wind power costs are projected to decline by as much as 30 percent by 2030.[8]

Low prices are setting in motion a rapid buildout of new power plant capacity. Wind installations have more than doubled since 2010, with more than 430 gigawatts installed worldwide. Meanwhile, global solar installations nearly quintupled between 2010 and 2015, reaching 227 gigawatts at the end of that period. Prices will decline even further as more capacity is built out, creating a virtuous cycle of more, cheaper clean power projects.

Battery storage has experienced similar cost declines. Within just a year and a half, the price of lithium-ion batteries has declined 70 percent, and prices are estimated to continue to drop nearly 50 percent in the next 5 years.[9] Flow batteries and new chemistry batteries are also expected to witness major cost reductions within that timeframe. Research suggests large-scale battery storage could grow to more than 7 gigawatts globally and cost as little as $230 per kilowatt-hour by 2030.[10]

As more renewable electricity capacity comes online, grid operators are becoming better and more adept at integrating it into the electricity system. In some regions, renewables already make up more than half of all generation. It is now possible to envision a future in which renewables make up 80 percent or more of electricity generation, enabling deep carbon reductions. No additional technological breakthroughs are needed to meet these penetration levels.

Deploying these technologies comes with other benefits as well. Renewables such as wind and solar are zero-emission sources, meaning they have the co-benefit of reducing local air pollutants, such as particulates and ozone. And because these sources use no fuel to generate electricity, they are essentially free to operate once installed, meaning electricity will continue to get cheaper.

Energy Efficiency
Innovation in energy efficiency continues as well. Well-constructed buildings today use a fraction of the energy that older buildings use, thanks to advances in lights, windows, insulation, and heating and cooling systems—all while maintaining or even improving comfort and energy reliability. Home appliances, industrial equipment, and vehicles have also become more efficient, providing the same or better service while using much less energy to operate.

The proliferation of light-emitting diode (LED) light bulbs is one of the most successful examples of innovation in energy efficiency. Since 2008, the efficiency of LEDs has approximately doubled, while prices have declined by 90 percent. New LEDs use about one-eighth as much energy as the incandescent bulbs they replace and last about 20 times longer. As a result, more than 80 million LED bulbs have been installed in the United States today, which have avoided millions of tons of CO_2 emissions and saved billions of dollars. Lighting is responsible for approximately 20 percent of the world's building sector electricity consumption,[11] meaning efficiency gains in this realm and others can add up to real energy and emission savings.

In all, the International Energy Agency (IEA) calculates that energy efficiency investments in IEA countries have saved 2,200 terawatt-hours of electricity—more than a tenth of global electricity consumption in 2015—avoiding more than 10 billion tons of CO_2 and saving \$550 billion in avoided energy costs.[12] By 2030, it is estimated that improved lighting efficiency could result in electricity savings equivalent to current electricity demand for all of Africa.

With these advances in renewable energy and energy efficiency technologies, many countries have decoupled or are beginning to decouple their energy use from economic productivity. Growth in the clean energy industry has created millions of new jobs, and prices for renewable energy sources and efficiency are now competing with (and often beating) prices for fossil fuel energy. A low-carbon future now costs the same as or less than a high-carbon one.

We Know Which Policies Can Achieve Effective Emission Reductions

Decades of energy policy examples have highlighted which policies are most effective in reducing carbon emissions and energy use. For instance, we know a strong building code that continuously strengthens over time and has a strong monitoring and enforcement mechanism, as in California, can dramatically reduce energy use and emissions. And we know fuel economy standards for vehicles, when designed well, can dramatically improve fuel efficiency.

We also now have resources that can help parse through the available policy options. The Energy Policy Simulator,[13] which allows users to evaluate the impact of hundreds of different climate and energy policies on emissions and costs, is one such tool. Another resource is the Clean Energy Solutions Center,[14] which connects policymakers with experts who can help craft effective climate and energy policy. We'll discuss these tools in greater detail in Chapter 3 ("How to Prioritize Policies for Emission Reduction").

The World Is Embracing These Technologies and Policies

The political will to enact strong climate and energy policies is stronger than ever. From local city ordinances to international treaties, politicians are lining up to put strong policies into action. At the international level, 189 countries have submitted emissions targets (intended nationally determined contributions [INDCs]) and signed the Paris climate treaty.[15] These commitments, which cover nearly 99 percent of the world's total emissions, are the first step in the battle to limit climate change. If met by all countries on time, the targets get the world about one-third of the way to the global goal of limiting climate change to tolerable levels (Figure I-8). Of course, pledges don't result in emission reductions; strong climate and energy policies with stringent monitoring

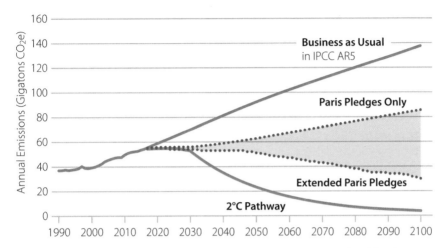

Figure I-8. Pledges made as part of the Paris Agreement get us partway to the 2°C pathway. (Graph data reproduced with permission from Climate Interactive and Climate Action Tracker, "Climate Action Tracker: Global Emissions Time Series," Climate Action Tracker, 2015, http://climateactiontracker.org/assets/Global/december_2015/CAT_public_data_emissions_pathways_Dec15.xls; "Scoreboard Science and Data," Climate Interactive, December 20, 2013, https://www.climateinteractive.org/programs/scoreboard/scoreboard-science-and-data/.)

and enforcement will be key to turning the commitments into real emission reductions.

The private sector is also embracing a low-carbon future. Businesses around the world are making ambitious pledges to cut their emissions. Nearly 180 companies, from Autodesk to Xerox, have signed the Science Based Targets initiative,[16] setting emission reduction targets. Beyond this pledge, 69 companies have joined the RE100 initiative,[17] committing to 100 percent renewable power, and many are more than halfway to meeting this target.[18]

Other public sector organizations, faith-based groups, foundations, and universities have also shown their support for curbing emissions by pulling their investments out of fossil fuels. In total, more than 550 institutions with assets of $3.4 trillion (a much smaller portion of which is invested in fossil fuels) have divested.[19]

Consumers, too, are shifting their behavior to reduce their carbon footprint. Households are installing solar panels (or, where that's not feasible, opting into green power programs offered by their utilities or joining community solar programs), buying energy-efficient appliances, and driving electric vehicles (EVs). For EVs specifically, more than a million have been sold around the world—reflecting breakneck growth rates from nearly zero just 5 years ago— and millions more are estimated to hit the roads in coming years as technology advances and production costs decline further.

In many cases, people take these actions purely because they are turning out to be cheaper than continuing their behaviors as usual. Smart energy devices such as thermostats and lighting systems are saving consumers money while improving comfort.

Development in low-emission technologies is providing a wide array of options for emission abatement. However, to have a shot at a reasonable climate future, policymakers will need to help push these technologies into the marketplace, and that requires smart policy. To figure out which policies can help achieve these goals, it is critical to quantify each major source of greenhouse gas emissions.

The Sources of Greenhouse Gas Emissions

There is no way to achieve a reasonable future unless the world focuses, first and most intensively, on the highest-potential abatement opportunities. The first step in this process is to identify the sources of greenhouse gas emissions around the world.

Nearly 75 percent of global greenhouse gas emissions are generated by just 20 countries (Figure I-9). Focusing efforts in these countries offers the highest potential for emission reductions.

Globally, and especially within the top 20 countries, emissions from energy combustion and industrial processes (which in this book include agriculture and waste) are the primary source of greenhouse gases, comprising more than 93 percent. Energy makes up nearly 74 percent of emissions, with industrial processes coming in at just under 20 percent. Targeting emissions from energy and industrial processes has the greatest potential for reductions.

Drilling down into the energy sector, emissions are fairly evenly spread across subsectors. The electricity sector is responsible for the largest share of emissions due to the combustion of coal, gas, oil, and biomass for power generation. Next up is industry, with emissions from energy combustion for electricity and heat. Transportation is third, followed by buildings. Of course, many

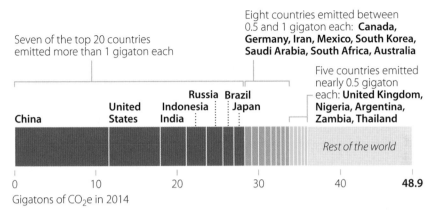

Figure I-9. The top 20 emitting countries are responsible for roughly 75 percent of global emissions. (Graph data reproduced with permission from CAIT Climate Data Explorer, 2017 [Washington, DC: World Resources Institute], available online at cait.wri.org.)

sectors are interrelated: The efficiency of building components determines the amount of electricity used by buildings, and the quality of building construction affects the amount of materials needed to fix existing buildings or build new ones, driving industrial energy use and emissions.

Industrial process emissions are dominated by emissions related to livestock, natural gas and petroleum systems, cement production, landfills, and refrigerants. Chapter 12 ("Industrial Process Emission Policies") includes a complete breakdown and a discussion of process emission policies.

This quantitative assessment points toward a straightforward roadmap for reducing energy-related emissions: Implement policies that reduce emissions in the electricity, industry, transportation, and building sectors in the top 20 emitting countries. For more information on targeting energy policy toward top sources of emissions, see Chapter 1 ("Putting Us on Track to a Low-Carbon Future").

Essential Climate and Energy Policy

For energy and climate policy to be effective, a suite of policies is needed; there is no silver bullet in this business. To design an optimal suite of policies, a policymaker should consider policies of four broad types: performance standards, economic signals, support for R&D, and enabling policies. Together, they create a powerful symbiosis that can drive deeper carbon emission reductions than policies in isolation while increasing cost-effectiveness.

- **Performance standards:** Performance standards set minimum requirements for energy efficiency, renewable energy uptake, or product performance. Examples include vehicle fuel economy standards, energy-efficient building codes, renewable portfolio standards, and power plant emission limits.
- **Economic signals:** Economic signals are policies designed to accelerate the adoption of clean energy technologies, ensure that positive and negative social impacts (i.e., externalities) are incorporated into product costs, or otherwise use the market as a tool for efficiently achieving emission reductions. Examples include carbon taxes and subsidies for clean energy production or efficiency upgrades.
- **Support for R&D:** Government support for R&D can accelerate innovation. New technology spurs economic development and reduces reliance on expensive and volatile fossil power sources. Government support can come in the form of funding for basic research (on technologies far from commercialization) intended to benefit many new industries. However, one of the most powerful ways government can support R&D is by creating an environment where private sector R&D can thrive. Examples include sharing technical expertise and facilities (such as national laboratories); adopting appropriate intellectual property protections; promoting robust science, technology, engineering, and mathematics education in public schools and universities; and structuring immigration laws so companies are not prevented from hiring foreign talent in science, technology, engineering, and math (STEM).
- **Enabling policies:** Enabling policies enhance the functionality of the other policies, often through direct government expenditures, information transparency, or reduction of barriers to better choices. For example, a policy requiring clear energy use labels on products allows consumers to make smarter decisions, and good urban design gives people transit options other than driving their car, enabling them to respond to well-designed economic signals.

When setting energy policy, it is best not to fall in love with a particular technology or a particular policy. For example, neoclassical economists love

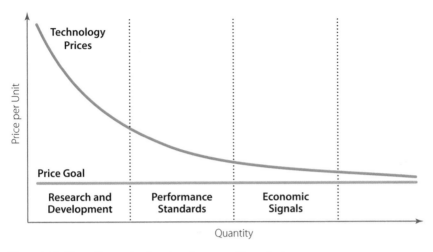

Figure I-10. The policy–technology learning curve (illustrative).

economic signals, such as carbon taxes. This is indeed the best policy for sectors that are highly price sensitive and for which there are low-carbon alternatives at near-competitive prices. But it may prove quite limited in other circumstances. R&D support wins the hearts of many a technologist, but this strategy does little to bend the practices of incumbent companies seeking to recover the costs of old technology, such as coal plants.

The beginning of wisdom in selecting a policy type is to map the technology, or the problem to be solved, on a learning curve, which shows how a technology price changes as the volume of production increases (Figure I-10).

Advanced nuclear power, carbon capture and sequestration, algal fuels, and dozens of other intriguing options require serious, sustained R&D. R&D, and innovation in general, has been one of America's greatest strengths, but we have shorted it, both in quantity and at times in management, so we do not have the shelf of options we should.

Once the basic principles of a technology are proven and early production is under way, a big and costly effort is needed to drive the price down to a reasonable level. We have seen this work dramatically in the last decade with wind and solar power, which experienced price drops of more than 80 percent and 60 percent, respectively. Over a slightly longer period, refrigerator energy consumption dropped by 80 percent. This improvement was realized because a number of jurisdictions set clear performance standards for those technologies. Today, offshore wind, central station solar, zero-net-energy buildings, and other technologies can improve most quickly if backed by clear performance standards. It's worth noting that performance standards are common and commonly accepted by even the most hardcore economists in many realms: Our building codes ensure that buildings don't easily burn or fall down, meat standards prevent innumerable cases of poisoning, and clean water is rightly considered a right.

Finally, for many sectors and technologies, pricing is the key. Removing subsidies for fossil fuels is the first step—though still widely ignored. Next, policymakers must incorporate the cost of externalities, such as adding a carefully derived social cost of carbon or setting a carbon cap.

Policy Design Principles

Of course, there is more to good policy design than just selecting the strongest set of policies. Each policy must be well designed in order to function as intended and achieve a policymaker's desired outcome—and a number of policy design principles can help. Careful application of these design principles, listed in Table I-1, can make the difference between a policy that works and a policy that fails. For more information on the four essential types of energy policy and these policy design principles, see Chapter 2 ("Energy Policy Design").

Together, targeted interventions in the top emitting countries using the best, well-designed policy options can put us on a path to a low-carbon future. The imperative to act rapidly and intelligently is overwhelming. This book offers a clear, workable strategy toward that.

Prioritizing Policies for Emission Reduction

Policymakers have many options available to them to reduce emissions from energy and industrial processes. The first step in evaluating which policies to prioritize is to assess the structure of the economy and emissions. Knowledge about how many cars, buildings, and power plants there are, how much energy they use, how their use is expected to grow over time, and so on can highlight which areas of the economy should be a focus for emission abatement.

The next step is to quantitatively evaluate the potential for policies to reduce emissions. Often this is done through marginal abatement cost curves (discussed in Chapter 3, "How to Prioritize Policies for Emission Reduction"),

Table I-1 Energy Policy Design Principles		
Performance Standards	**Economic Signals**	**Support for R&D**
Create long-term certainty of the standards to provide businesses with a fair planning horizon.	Create a long-term goal and provide business certainty.	Create long-term commitments for research success.
Build in continuous improvement.	Price in the full value of all negative externalities for each technology or	Use peer review to help set research priorities.
Focus standards on outcomes, not technologies.	Use a price-finding mechanism.	Use stage-gating to shut down underperforming projects.
Prevent gaming via simplicity and avoiding loopholes.	Eliminate unnecessary soft costs.	Concentrate R&D by type or subject to build critical mass.
	Reward production, not investment, for clean energy technologies.	Make high-quality public sector facilities and expertise available to private firms.
	Capture 100% of the market and go upstream or to a pinch point when possible.	Protect intellectual property without stymieing innovation.
	Ensure economic incentives are liquid.	Ensure that companies have access to high-level STEM talent.

which look at the ability and cost-effectiveness of specific *technologies* to re-duce emissions. A new advance, and improvement, is to instead evaluate the ability and cost-effectiveness of specific *policies* to reduce emissions, through policy cost curves. An analysis grounded in policy rather than technology gives policymakers a more direct path to emission reductions because it taps into their ability to implement policy.

Not every country needs to conduct all these analyses. Some regions will be quite similar to others that have already conducted some or all of these assessments, in which case similar findings will apply.

Additionally, decades of experience in policy design backed up by quantitative computer modeling demonstrate that a small set of these policies (Table I-2), designed well, can effectively reduce emissions. Each of these policies is introduced and discussed in detail in Part II of this book.

In the power sector, renewable portfolio standards and feed-in tariffs can reduce emissions by increasing the share of fossil-free power generation.

Table I-2 — Most Effective Policies for Reducing Emissions				
Power Sector	**Transportation Sector**	**Building Sector**	**Industry Sector**	**Cross-Sector**
Renewable portfolio standards and feed-in tariffs Complementary power sector policies	Vehicle performance standards Vehicle and fuel fees and feebates Electric vehicle policies Urban mobility policies	Building codes and appliance standards	Industrial energy efficiency policies Industrial process emission policies	Carbon pricing R&D policies

Designed well, they can minimize the costs of transitioning to a low-carbon power system. Complementary policies, such as support for transmission, smart utility policy design, and efficiency resource standards, are important as well.

Strong standards and incentives to improve energy use in industry can significantly reduce emissions from the industrial sector. The industry sector also has significant process emissions, which include emissions of CO_2 and other non-CO_2 gases generated in industrial processes. Policymakers can require control of these emissions and offer incentives and other forms of assistance to encourage reductions.

In the building sector, building codes and appliance standards are the best tools for reducing emissions. These policies tend to save money as well, because over time decreased energy use outweighs any increase in costs.

Fuel economy standards, vehicle feebates, electric vehicle incentives, and smart urban planning can all reduce emissions significantly in the transportation sector by increasing the fuel efficiency of vehicles, decreasing their emissions, and offering alternative transportation options.

Carbon pricing is another strong tool for reducing emissions, encouraging emission-reducing behavior across the economy and pushing investments to lower-carbon options.

Finally, support for R&D helps reduce the costs of all of these policies while providing opportunities for new low-carbon technologies to hit the market.

For more on which policies can most effectively drive down emissions, see

Chapter 3 ("How to Prioritize Policies for Emission Reduction") and individual chapters in Part II of this book.

How to Use This Book

This book should be used as a resource by policymakers, advocates, philanthropists, and others in the climate and energy community as a guide to where to focus efforts and how to ensure that policy is designed to maximize success. Part I of the book provides readers with a roadmap for understanding which countries, sectors, and sources produce the greatest amount of greenhouse gas emissions. It should help readers understand that focusing on energy and industrial process emissions in the highest-emitting countries is the most effective way to reduce greenhouse gas emissions. Part I also provides readers with insight into how they can choose which policies to focus on and which to prioritize.

In Part II of the book, readers can explore each of the top emission-reducing policies in detail. Each chapter includes detailed information on the policy and its goals, when to apply each policy, the key policy design principles that make that policy effective, and case studies of good and bad applications of that policy. At the end of a chapter on a specific policy, readers will be able to identify the right situations in which to apply that policy and the key design elements that are necessary.

Conclusion

Climate change requires action as soon as possible to limit emissions and avoid exceeding two degrees of warming. Governments, businesses, and organizations around the world have committed to reducing emissions, laying the foundation for deeper emission cuts that put the world on a trajectory to a lower-carbon future. The key now is in turning these pledges into reality—with laser-focused, well-designed policy.

This task is by no means impossible. We have the technology today to rapidly move to a clean energy system. And the price of that future, *without counting environmental benefits*, is about the same as that of a carbon-intensive future. So the challenge is not technical, nor even economic, but rather is a matter of enacting the right policies and ensuring they are properly designed and enforced.

PART I

A Roadmap for Reducing Greenhouse Gas Emissions

Significant reductions in greenhouse gas emissions are necessary to limit climate change and stay under two degrees of warming by the end of century. To identify where reductions will be most effective, it is necessary to examine the greatest sources of emissions. Seventy-five percent of global greenhouse gas emissions come from just 20 countries. Furthermore, 94 percent of emissions come from industrial processes and the energy used in power and heat generation, transportation, industry, and buildings. These findings show that we must focus our efforts on reducing emissions from industrial processes and energy in the top 20 countries.

Policymakers have many options available to tackle emissions. Policies can generally be classified as one of four types, each of which reinforces the others: performance standards, economic signals, support for research and development (R&D), and supporting policies.

Performance standards improve new equipment and help capture savings that economic signals cannot, because of market barriers. **Economic signals** can be highly efficient and encourage the uptake of more efficient equipment driven by performance standards. Support for **R&D** and **enabling policies** lower the costs of performance standards and economic signals by pushing new technologies to market and lowering the costs of existing technologies by removing deployment market barriers.

There is no silver bullet policy for tackling emissions. Rather, a portfolio of policies that reinforce one another is the best approach. In Part I, we'll discuss the suite of policies available, strategies for identifying the most effective options, and principles for designing successful policy programs. Quantitative modeling reveals that these policies, designed and implemented well, can put the world on track to a future where warming is limited to two degrees or less.

CHAPTER ONE

Putting Us on Track to a Low-Carbon Future

As outlined in the Introduction, significant reductions in greenhouse gas emissions are needed to avoid the worst impacts of climate change. But how much effort is needed? What types of reductions and emissions pathways are needed in order to avoid the worst parts of climate change? And how we do know where to focus our efforts? This chapter tackles these questions and highlights the sectors where our efforts will have the greatest impact.

Avoiding the Worst Impacts of Climate Change

The level of greenhouse gases in the atmosphere is measured in parts per million, or the number of greenhouse gas particles per million particles in the atmosphere. The impact of gases other than carbon dioxide is measured by equating those gases to an equivalent amount of carbon dioxide, called carbon dioxide equivalent (CO_2e). The equivalence of gases ranges widely. For example, 1 molecule of methane equals about 30 molecules of carbon dioxide, whereas other chemicals such as fluorinated gases, used primarily as refrigerants, are thousands of times more potent than carbon dioxide per molecule. Notably, the equivalence value varies based on the timeframe over which the gas is evaluated (methane has a higher equivalence over 20 years than over 100 years, for example) and as the science of climate change advances. The total amount of CO_2e in the atmosphere includes CO_2 as well as all the other gases that contribute to climate change.

There is broad consensus that preventing the worst impacts of climate change requires keeping global warming below two degrees Celsius through the end of the 21st century. To have at least a 50/50 chance of limiting warming to two degrees, we must limit concentrations of CO_2e to 500 parts per million by 2100, although some overshoot of this target in previous years is okay.[1] Yet in 2015, CO_2e concentrations measured 485 parts per million, and they have been increasing at a rate of 2–4 parts per million per year.[2] To achieve the 500 parts per million target by 2100, immediate on-the-ground action is needed. But what does this mean in terms of emissions?

Climate change and the warming that drives it are a function of the total amount of carbon in the atmosphere. In other words, it is a stock problem, not a flow problem, as discussed in the Introduction. Therefore, it is useful to think of emissions, and necessary emission reductions, in terms of cumulative totals rather than annual amounts. Significant action to reduce emissions will be needed throughout the 21st century, but for simplicity and given the growing uncertainty in years further out, we focus on the necessary reductions between now and 2050.

Without additional action to reduce greenhouse gas emissions, just over 2 trillion tons of CO_2e will be emitted between 2016 and 2050.[3] Although climate models vary, they show that in order to meet the 500 parts per million target, cumulative total emission reductions of 25 to 55 percent relative to a business-as-usual scenario are necessary between 2016 and 2050.[4]

For this book, we rely on modeling completed in 2013 as part of the Low Climate Impact Scenarios and the Implications of Required Tight Emissions Control Strategies (LIMITS) exercise. In particular we rely on the modeling done by Pacific Northwest National Laboratory and the Joint Global Change Research Institute using the Global Change Assessment Model, evaluating emissions between 2010 and 2050. More information on the Global Change Assessment Model, the LIMITS study, and emission scenarios from the Inter-governmental Panel on Climate Change is provided in Appendix II.

The results of the LIMITS study suggest that to have a 50/50 shot at staying under two degrees of warming we need to reduce cumulative greenhouse gas emissions by at least 41 percent between 2010 and 2050 (Figure 1-1).

This value is global; emission reductions needed from individual countries will vary, depending on their development status. For example, the most in-dustrialized countries will need to achieve significantly deeper reductions than the 41 percent global number to compensate for other emerging economies with high rates of economic development. It's also worth noting that a 41 per-cent reduction in *cumulative* emissions entails much greater *annual* emission reductions in later years as emission reductions are phased in. In 2050, global annual emission reductions of 65 percent relative to business-as-usual will be necessary, with the more economically developed regions needing to achieve reductions of 70 percent or more.

This book evaluates potential reductions at a global scale. According to the Global Change Assessment Model results discussed earlier, we need cumulative greenhouse gas emission reductions of just over 40 percent between 2020 and 2050 relative to business as usual to give ourselves a 50/50 shot at staying under two degrees of warming. This is the target we aim for in this book.

The Paris Agreement: A Good First Step

In December 2015, 189 countries responsible for nearly 99 percent of the world's greenhouse gas emissions signed the Paris Agreement,[5] in which they agreed to make an effort to limit emissions over the next 10 to 30 years. The centerpiece of the Paris Agreement is each country's specific emission reductions targets.

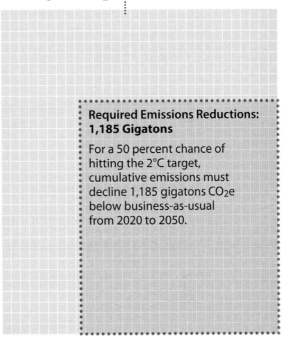

Business As Usual Cumulative Emissions, 2020-2050: 2,253 Gigatons CO$_2$e

Required Emissions Reductions: 1,185 Gigatons

For a 50 percent chance of hitting the 2°C target, cumulative emissions must decline 1,185 gigatons CO$_2$e below business-as-usual from 2020 to 2050.

Figure 1-1. Emission reductions needed for a 50 percent chance of avoiding 2°C global warming. (Analysis done using data with permission from the International Institute for Applied Systems Analysis [IIASA]. Data source: Tavoni et al., 2013. Data downloaded from the LIMITS Scenario database hosted at IIASA, https://tntcat.iiasa.ac.at/LIMITSPUBLICDB/dsd?Action=htmlpage&page=about.)

If the targets are all met, they would collectively move emissions a good share of the way to the two-degree pathway. As shown in Figure 1-2, the Paris Agreement commitments, on their own, move the emission curve about a third of the way to the two-degree pathway relative to business-as-usual. If existing policies and the Paris pledges are extended to 2100 with the same degree of effort, the emission curve moves about 80 percent of the way to the two-degree pathway. Despite the United States' decision to withdraw from the Paris Agreement, commitments from remaining countries still cover more than 80 percent of the world's emissions today. Furthermore, U.S. states, cities, and

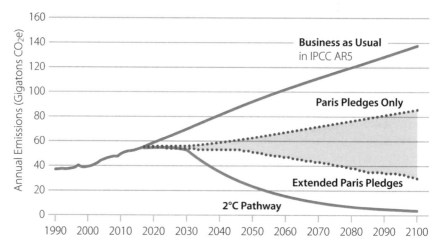

Figure 1-2. Pledges made as part of the Paris Agreement get us partway to the 2°C pathway. (Graph data reproduced with permission from Climate Interactive and Climate Action Tracker, "Climate Action Tracker: Global Emissions Time Series," Climate Action Tracker, 2015, http://climateactiontracker.org/assets/Global/december_2015/CAT_public_data_emissions_pathways_Dec15.xls; "Scoreboard Science and Data," Climate Interactive, December 20, 2013, https://www.climateinteractive.org/programs/scoreboard/scoreboard-science-and-data/.)

businesses have expressed their commitment to meeting emission reduction targets, which will help reduce U.S. emissions.

The commitments enshrined in the Paris Agreement represent a significant diplomatic accomplishment and provide a very important impetus to move the global economy in the right direction. However, the existing commitments do not themselves add up to the two-degree pathway. And, perhaps more importantly, *the pledges on their own will not result in on-the-ground emission reductions. Domestic policy is needed to drive change in the power plants, factories, buildings, vehicles, and forests.* These shortcomings raise two important questions: First, how can policymakers close the gap between the existing Paris commitments and the two-degree pathway? Second, how can policymakers translate the targets into real-world emission reductions?

Focus on the Highest-Emitting Countries

Although the Paris Agreement encompasses nearly 99 percent of global emissions (not including the proposed U.S. withdrawal, which drops it down to

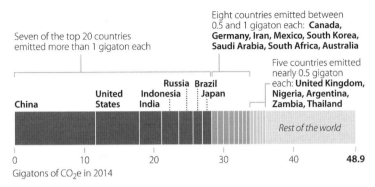

Figure 1-3. The top 20 emitting countries are responsible for roughly 75 percent of global emissions. (Graph data reproduced with permission from CAIT Climate Data Explorer, 2017 [Washington, DC: World Resources Institute], available online at cait.wri.org.)

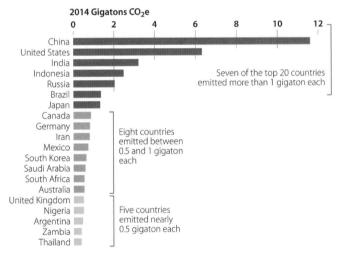

Figure 1-4. Emissions vary widely in the top 20 emitting countries, dominated by China and the United States. (Graph data reproduced with permission from CAIT Climate Data Explorer, 2017 [Washington, DC: World Resources Institute], cait.wri.org.)

about 82 percent), just 20 countries account for nearly 75 percent of global greenhouse gas emissions (Figures 1-3 and 1-4). The top 20 emitting countries all submitted pledges in 2015 (although, as noted, the United States has announced its withdrawal since then), but many of these countries have the potential to significantly strengthen their commitments. For example, Climate

Action Tracker, an independent group that tracks and evaluates climate policy, rates the following countries' pledges as "Inadequate": Russia (4th largest emitter), Indonesia (5th largest emitter), Japan (7th largest emitter), Canada (8th largest emitter), Australia (12th largest emitter), South Korea (13th largest emitter), and South Africa (17th largest emitter).[6] Even the two largest emitters, China and the United States, have only "Medium" ratings for their pledges.[7] The weak contributions from many of the top-emitting countries, including 4 of the top 10, suggest that targeting these countries for additional reductions could make a positive impact on global emissions and help move global commitments closer to the two-degree pathway.

Energy and Industrial Processes Drive Greenhouse Gas Emissions

The second and perhaps more important question is: How do countries translate pledges, which are simply high-level emission targets, into actionable policy that will achieve real-world emission reductions? Answering this question requires an assessment of what sources are responsible for greenhouse gas pollution.

Energy and industrial process (including agriculture and waste) emissions are by far the largest driver of CO_2e emissions globally (Figure 1-5).[8] Energy-related emissions account for just under 74 percent of global emissions, and industrial processes account for nearly 20 percent. Together, they total nearly 94 percent of global greenhouse gas emissions. In some countries, such as Indonesia, Brazil, and Nigeria, deforestation and other land use change emissions are significant sources of greenhouse gases.[9]

The sources of industrial process

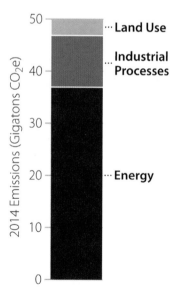

Figure 1-5. CO_2e emissions are primarily from energy and industrial processes. (Graph data reproduced with permission from CAIT Climate Data Explorer, 2017, [Washington, DC: World Resources Institute], cait.wri.org.)

emissions are well documented, and specific policies targeting those emissions are discussed in Chapter 12 ("Industrial Process Emission Policies"). Given the fact that energy is the largest source of greenhouse gas emissions, the next logical question becomes: What drives energy-related greenhouse gas emissions?

Energy-related greenhouse gas emissions are concentrated in the electricity,[10] transportation, building, and industry sectors, with power plants generally being the largest source (Figure 1-6).[11] These emissions come primarily from burning coal and natural gas to create power and heat and burning petroleum products to power vehicles.

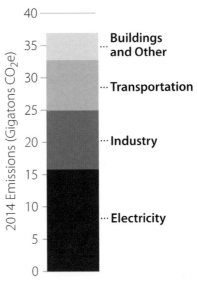

Figure 1-6. Energy and industrial process CO_2 emissions are fairly evenly spread across the electricity, industry, transportation, and building sectors. (Graph data reproduced with permission from CAIT Climate Data Explorer, 2017 [Washington, DC: World Resources Institute], cait.wri.org.)

A Roadmap to a Low-Carbon Future: Focus on the Biggest Sources in the Top Countries

The sector-by-sector math of CO_2e emissions, overlaying the 20 countries that are the largest sources, sheds light on where to focus efforts. **Quite literally, there is no path to a low-carbon future other than the list below. Every policy idea must be measured against its contribution to one or more of these goals.**

Reduce Electricity Demand in the Building and Industry Sectors
Demand for electricity is driven by buildings and industry, and increasing their efficiency is a large-scale, cost-effective strategy. Efficiency is typically the most cost-effective way of reducing emissions, with initial investments paying dividends for years via reduced fuel costs.

Reduce the Carbon Intensity of Electricity Generation
Electricity sector emissions can also be lowered by reducing the carbon intensity of electricity generation. Using fossil-free technologies such as wind, solar,

hydro, geothermal, and nuclear to generate electricity can avoid the emissions (and also the air quality problems) that come from burning fossil fuels such as coal and natural gas.

Reduce Transportation Emissions through Efficiency, Electrification, and Urban Mobility

The transportation sector is a large and growing source of greenhouse gas emissions. The top ways to reduce pollution from transportation are to improve vehicle fuel economy, electrify vehicles (and to simultaneously reduce the carbon intensity of electricity generation), and provide alternatives to personal vehicle travel via smart urban planning and public transit.

Reduce Non–Electricity Industry Sector Emissions

Non–electricity industry sector emissions are another large source of greenhouse gas emissions. These include primarily industrial process emissions (e.g., the chemical processes involved in cement manufacturing or natural gas venting and flaring) but also energy used for heat, as in the iron and steel industry.

Reduce Deforestation and Forest Degradation in Tropical Forest Nations

In tropical forest nations where a large share of emissions come from land use, land use change, and forestry, policymakers should aim to reduce deforestation and forest degradation. A handful of options exist to achieve these goals,[12] including legally protecting forests through the creation of designated protected areas, payments to landowners for providing ecosystem services, and payments to landowners to remove forested land from timber production.

Although land use is an important sector for emission reductions, this book focuses on energy and industrial process emission reductions. The science, the policies, and the actors for reducing emissions from land use are very different from those for energy and industrial processes, and they deserve separate treatment from experts in land use policy.

Conclusion

The Paris Agreement targets, if fully achieved, get us about one-third of the way to the two-degree goal, meaning further reductions will be necessary. But more important, the commitments under the Paris Agreement are targets, and unless they are converted into highly effective, sector-specific national policies, they will achieve little. The aim of this book is to help guide that process.

The starting point is to evaluate where emissions are coming from. Energy and industrial processes are the dominant sources of greenhouse gas emissions in most economies. Within the energy sector, emissions are evenly spread across the electricity, industry, transportation, and building sectors. This assessment suggests that to reduce emissions, policymakers need to focus on reducing electricity demand in the industry and building sectors, reducing the carbon intensity of electricity generation, improving the efficiency of vehicles while providing cleaner alternatives, and reducing process emissions in the industry sector. In certain economies, a strong focus on reducing emissions from land use change is also necessary.

Now that we know what we need to do reduce emissions, the next question is: How do we achieve these goals? To examine this question, we turn next to the four essential types of energy policy.

CHAPTER TWO

Energy Policy Design

We have now evaluated the size of reductions necessary to put us on a path to meeting the two-degree target and examined the key sources of greenhouse gas emissions. Reducing energy-related emissions from electricity, buildings, transportation, and industry and industrial process emissions is the only way to achieve deep decarbonization.

But how can policymakers target reductions from these sources using policy? To answer this question, it is first important to understand the four types of essential energy policy and how they reinforce and interact with one another.

Essential Energy Policy

Many policymakers understand the urgent need to reduce greenhouse gases and mitigate the worst impacts of climate change, but they need data to help sort through the many types of policies that can help. Different policies are best suited for different circumstances, and some policies that look good on paper fail to perform in the real world. Despite this complexity, a practical consensus about what works is emerging, and it combines performance standards, economic signals, and research and development (R&D). In addition to these fundamental policy types, other enabling policies are also necessary, such as strategies that can lower the financial risk for deploying emerging low- and zero-emission technologies.

It is worth making the point here again that *there is no silver bullet policy for dealing with climate change.* Many economists will argue that carbon pricing is a panacea and can drive all the change that is needed. This is false. Carbon pricing, a type of economic signal, is useful and effective in many settings, but it is ineffective in parts of the economy prone to market failures. The limitations of carbon pricing in particular are introduced in this chapter and discussed in detail in Chapter 13 ("Carbon Pricing"), but in short, market failures often result in carbon pricing not delivering its expected impacts.

This leads to the next point worth stating bluntly: *A portfolio of policies, including performance standards, economic signals, support for R&D, and supporting policies, is the most effective, lowest-cost way to drive down greenhouse gas emissions.* Properly designed, they reinforce each other through system dynamics that emerge organically. However, a portfolio is not simply a grab-bag of policies. There are hundreds of policy options that have little value. The right policies must be selected for each sector, and then they must be designed and implemented well.

Performance Standards

Performance standards set quantitative targets at the device, fuel, or sector level; they specify what level of performance businesses or equipment must achieve, such as fuel economy standards for vehicles or particulate emissions standards for coal power plants.

Performance standards provide a quantity signal or minimum performance criteria where price is not an effective inducement. They help set guardrails for the market, allowing competition within those guardrails, which favors least-cost solutions that meet the constraint.

Performance standards are particularly valuable for motivating low-cost energy efficiency reductions that are not price responsive. For example, consumers are often uninterested in a more efficient appliance or vehicle unless any increased upfront price will pay itself back in fuel savings in less than 1 or 2 years (i.e., they have a very high discount rate). Performance standards increase the availability and uptake of price-competitive efficient and low-carbon options, making them consumer ready.

Another role for performance standards is to spur technological innovation essential to long-run decarbonization. Without a demand signal stimulated by performance standards, there may be insufficient investment opportunities from private companies to fund research and development in new areas.[1] However, with strong performance standards and a clear timeline over which they will become more stringent, companies have a strong incentive to invest in innovation.

Take the case of electric vehicles (EVs), which are widely recognized as being a linchpin of global efforts to decarbonize.[2] Performance standards to push EV deployment and innovation forward are needed because the near-term technologies will not be the lowest-cost options for auto fuel efficiency. In this instance, performance standards may be somewhat more expensive per unit carbon saved than economic signals, but they are needed in the short term to create the conditions for low-cost options in the long run. Examples of performance standards for this task include a zero-emission vehicle (ZEV) standard (requiring manufacturers to sell ZEVs as some fraction of total sales) or a low-carbon fuel standard (where the lifecycle carbon intensity of fuel decreases over time). The second type of case for performance standards is when market barriers have limited the adoption of energy efficiency technologies, even when there is a clear economic rationale for these technologies.

Performance standards have limitations (as do economic signals and R&D). Many performance standards target only new products, which is a particular limitation for long-lived investments such as building heating and cooling systems. They require regulators to be knowledgeable about the technology and business operations they are targeting. Performance standards must be stringent enough to force energy innovation but must be reasonable in terms of cost effectiveness and of what manufacturers will realistically be able to accomplish.

Economic Signals

Economic signals for climate policy come in two flavors: fees that discourage pollution, such as a carbon tax, and subsidies that encourage cleaner alternatives, such as incentives for energy-efficient products.

In terms of emission reduction, the most widely discussed policy is carbon pricing, which is covered in Chapter 13 ("Carbon Pricing"). Carbon pricing creates a signal that radiates across all sectors of the economy, affecting both the purchase of goods and their use. It is technology neutral and generates an efficient source of revenue, which can be helpful for accomplishing other policy objectives.

Many other economic signals are important as well. For example, power sector feed-in tariffs—which provide power plants with a fixed payment for each unit of electricity they generate and are covered in Chapter 4, "Renewable Portfolio Standards and Feed-In Tariffs," can help decarbonize electricity. Transportation sector vehicle feebates, fees on inefficient vehicles rebated to purchasers of efficient ones and covered in Chapter 7, "Vehicle and Fuel Fees and Feebates"—encourage the purchase of more efficient vehicles.

Broadly speaking, economic signals are a helpful strategy for reducing emissions, but they are not a *sufficient* strategy for either short-run efficiency or long-run innovation. For example, well-known market failures and transaction barriers restrict the ability of economic signals to encourage adoption of low-cost—or even cost-saving—energy efficiency upgrades that would reduce emissions. There are many examples of these barriers, including split incentives, short payback horizons and inconsistent financial valuation, lack of up-front capital for investment, and a failure of the investor to capture the benefits of the investment.[3] For these reasons, economic signals are often best when paired with performance standards that can target emissions from sources with significant market barriers. A successful portfolio of policies will include both performance standards and economic signals.

Rental properties offer a good example of the need for this combined approach. Split incentives occur in most rental properties when renters, and not building owners, who typically make the capital investment decisions that affect energy efficiency, pay the energy bills. A landlord not paying utility bills has little reason to upgrade an apartment fitted with an inefficient water heater and refrigerator, but the renter is in no position to make a capital improvement on the building. The economic opportunity is missed, and economic signals alone won't fix it. In contrast, a good building code (a performance standard), properly enforced, can get the job done.

Although economic signals may reduce emissions less than in a completely efficient market because of market barriers, economic signals—particularly fees or taxes on pollution—can create new government revenue that can be reinvested in clean technologies and for other social causes. As a method of raising revenue to invest in low-carbon technologies, economic signals can be very effective.

R&D Support

Clean technology provides valuable environmental, health, and economic benefits that are not all represented, at least in the short term, in the prices people pay in the marketplace. Similarly, technological advances from investment in R&D lower the cost of future emission abatement and can therefore decrease the cost of any policy portfolio.

These uncaptured spillover effects from R&D create the need for policy support, which can be direct (e.g., government funding of research at universities or national laboratories) or may involve creating a policy environment that is favorable to private institutions doing their own research.

The key to deploying any technology is to achieve a decline in unit costs, which happens over the lifetime of a technology through learning in research settings, learning by doing in production and application, and economies of scale.

First, these price declines are driven by laboratory R&D, inventing and testing many iterations of a new technology. Then, in preparation for commercialization, a demonstration phase is necessary where engineering improvements help drive prices lower. As more units of the technology are deployed, price declines are driven by economies of scale and learning by doing (i.e., learning lessons about how to deploy the technology at scale and within a system, lessons that could not be learned solely from working with a new technology in a laboratory). Finally, once commercial viability is achieved and with increasing market penetration, decreasing production costs follow from large-volume economies of scale and additional learning by doing. Price declines are not automatic: A technology must be actively researched (in early stages), then incrementally improved and deployed (in middle and later stages), to realize cost reductions.

Learning curves is the term used to describe these regular patterns of declining costs in new technologies. Information technology is famous for having exponential learning curves. Energy technologies have also exhibited regular patterns of improvement in performance and cost.[4]

Solar photovoltaic (PV) electricity generation is a good example. This technology dates to the 1950s, but for many years it was too expensive to be used commercially, except in very limited circumstances, such as to power satellites. The price per watt of crystalline silicon solar cells was $76.67 in 1977.[5] Over time, laboratory research (including learning from the commercial semiconductor industry) drove down prices, and as prices declined, more commercial applications for solar became feasible and deployment accelerated. This started a feedback loop that further drove down prices. The price per watt of a solar cell reached $0.26 in 2016, a decline of 99.6 percent in 39 years.[6]

The learning curve in Figure 2-1 shows that every time the world's solar power has doubled, the cost of solar panels has fallen about 22 percent. Learning curves can help us select policy in combination with future emission targets. Once the emission reduction needs are known, the potential for reductions using currently available commercialized technologies can be evaluated. Once the remaining emission reduction deficit is known, emerging technologies can

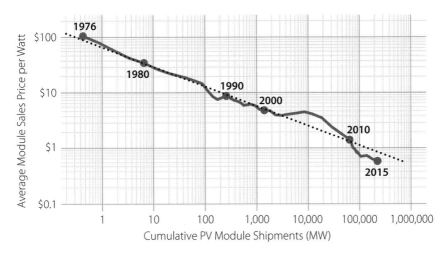

Figure 2-1. The costs of solar photovoltaic modules have decreased apace with deployment. (Diagram is a compilation of data according to references 1, 4, 5, 6, 7, 8 as mentioned on p. 48 of "International Technology Roadmap for Photovoltaic ITRPV," ITRPV; VDMA, 2017.)

be assessed, sector by sector, to find the most promising ones for R&D support, based in part on their position on the learning curve.

Working backward with an understanding of where we need to end up, policymakers can identify the most promising technologies. Of course, no one has a crystal ball. Some failures are inevitable in fundamental and early-stage research, but that does not make it any less essential for governments to support R&D for promising technologies.

Together with performance standards and economic signals, R&D support can help push laboratory technologies to market and provide new options for reducing emissions at low cost (Figure 2-2).

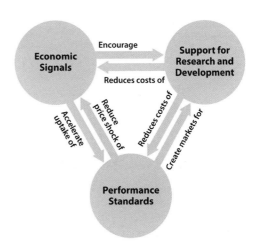

Figure 2-2. Different policy types drive and reinforce each other.

Supporting Policies

In addition to the three types of policies outlined earlier, a large group of supporting policies can improve the effectiveness and lower the cost of performance standards, economic signals, and R&D. Supporting policies tend to lower transaction costs, improve information, and streamline decision making.

For example, appliance efficiency labels are a type of supporting policy that doesn't set minimum performance criteria or price in externalities but provides purchasers with better information that can push them to choose the most efficient goods.

There are many different supporting policies, some of which are included in the policy-specific chapters in Part II of this book. However, these policies tend to be supportive of and secondary to the three types of policies discussed earlier.

Table 2-1 Energy Policy Design Principles		
Performance Standards	**Economic Signals**	**Support for R&D**
Create long-term certainty of the standards to provide businesses with a fair planning horizon.	Create a long-term goal and provide business certainty.	Create long-term commitments for research success.
Build in continuous improvement.	Price in the full value of all negative externalities for each technology or	Use peer review to help set research priorities.
Focus standards on outcomes, not technologies.	Use a price-finding mechanism.	Use stage-gating to shut down underperforming projects.
Prevent gaming via simplicity and avoiding loopholes.	Eliminate unnecessary soft costs.	Concentrate R&D by type or subject to build critical mass.
	Reward production, not investment, for clean energy technologies.	Make high-quality public sector facilities and expertise available to private firms.
	Capture 100% of the market and go upstream or to a pinch point when possible.	Protect intellectual property without stymieing innovation.
	Ensure economic incentives are liquid.	Ensure that companies have access to high-level STEM talent.

Policy Design Principles

Selecting between the four policy types discussed earlier is just the first step to a strong policy portfolio; the specific policies selected must be designed well to function effectively. Each type of policy has certain characteristics that determine whether it is successful. Experience with designing and evaluating energy policy over many years has elucidated a set of policy design principles that are essential components of performance standards, economic signals, and R&D policy, and that separate successful policies from failures (although supporting policies are an important element of any policy portfolio, their heterogeneity means they lack a set of design principles and are therefore excluded from this section).

In the next section, we'll identify the most important design principles for each type of policy. They are generally applicable based on policy type, although each principle will not apply to every policy. They are not overly complicated. In fact, most of these policy design principles are obvious, yet they continue to be overlooked in energy policy, often with disastrous consequences.

Table 2-1 includes a list of the policy design principles.

Performance Standards

Create Long-Term Certainty to Provide Businesses with a Fair Planning Horizon

Performance standards affect decisions and investments made by businesses. One firm might invest in R&D to improve energy efficiency and comply with an appliance standard, and another may finance a wind farm to meet a renewable portfolio standard, which requires a minimum share of electricity generation from renewable resources.

Business investments take time and come with a degree of risk. For example, it may take an auto manufacturer years, and cost tens or hundreds of millions of dollars, to succeed in research that drives markedly higher efficiency. Without long-term policy certainty, it might be too risky for the manufacturer to invest in that research. The manufacturer may believe there is a low chance of succeeding within the necessary timeframe or may not want to risk the upfront investment if they believe the policy has a high likelihood of changing before research investments pay off.

If a performance standard takes effect too soon after its adoption, businesses may be unable to meet the standard as cheaply or with products of as high a quality as would have been possible if they'd had a few more years to prepare.

Just as businesses should not be hit suddenly with tighter standards than they were expecting, performance standards should not be delayed or relaxed at the last minute. To delay or weaken the standards has the effect of benefiting laggards (companies that did not sufficiently invest in R&D and other preparations to meet the standard) while hurting companies that made the necessary investments to meet the standards in good faith. Companies that worked to meet the new standards rely on the standards to get a return on their R&D investment, because they will be able to offer appealing products that meet the standards at lower cost than competitors. If they are unexpectedly forced to compete with competitors' older, inefficient, but cheap products, they may not see the financial rewards they were expecting.

As a rule, in a given industry, standards should be known at least as many years ahead as it takes to complete a full product revision (that is, developing a new product, retooling factories to produce it, updating marketing materials, etc.). This timeframe varies by industry but generally is at least several years and may be as long as 10 years. Knowing standards even further ahead (say, two or three product revisions) might be even better.

Because of the uncertainty inherent in technological progress, it may not be possible to strictly specify what performance level is reasonable for a given technology to be required to meet 10 years in the future. There are two solutions to this problem.

The first solution is to base the new standard on a subset of the products available in the marketplace. For example, a standard might be set to the level achieved by the highest quintile of existing products, say, 3 years prior. This guarantees the standard can be met with available technology, because there existed commercialized products 3 years ago that met the standard.

The second option is to set a known schedule of when performance improvements will take effect, extending many years into the future, but the magnitude of the improvements is set closer to the date when they go into effect by a regulatory agency, with input from affected industries. This approach has the benefit of increased flexibility, but it reduces the certainty associated with a schedule of known improvements, and it makes the performance standard vulnerable to political interference and regulatory capture. Therefore, it is best used only when necessary to achieve consensus for tighter standards.

Build In Continuous Improvement

A performance standard must have a mechanism for automatic tightening, so it does not become stagnant and ineffective. This might continue until the technology starts to approach fundamental limits (e.g., theoretical efficiency limits imposed by thermodynamics), is replaced by another technology entirely (e.g., gasoline cars may be replaced by electric cars), or begins to saturate the sector (e.g., renewable electricity as a share of total electricity generation).

It is valuable for the improvement mechanism to be built into the law or regulation promulgating the standard. This means that the law specifies a requirement for tightening the standard over time. Adjustments to the standard may happen according to a known schedule (say, every 3 or 4 years), and the law may specify the amount of the adjustment or, if uncertain, the manner in which adjustments are to be determined (see the preceding section for two ways to do this). Alternatively, a standard may simply require a fixed annual percentage improvement.

Building improvement into the standard helps prevent long periods of stagnation that can result when legislative bodies or even regulatory agencies must opt to tighten the standard with legislation or rulemaking. In the United States, the fuel economy standard for automobiles reached 27.5 miles per gallon in 1985 and did not rise above this level until 2011, costing the United States hundreds of billions of dollars.

Japan's Top Runner program has an excellent mechanism for continuous improvement. The Top Runner energy efficiency standards cover many different categories of products, each reviewed on a known schedule. At the time of review, the efficiency of the most efficient product on the market is selected as the standard that must be met by all manufacturers within several years. The policy also accounts for additional potential technological improvements by slightly increasing the standard above the value of the most efficient product. To increase flexibility, manufacturers must meet the standard on the basis of weighted average of all their shipments of products in a given category. A manufacturer might comply by selling only products that meet the standard, or a manufacturer might offer a product that fails to meet the standard, as long as the manufacturer also sells enough products that exceed the standard to bring the weighted average efficiency of all shipments above the standard. As a result of the Top Runner program, passenger vehicles improved in efficiency by 49 percent from 1995 to 2010, refrigerators by 43 percent from 2005 to 2010, and TV sets by 61 percent from 2008 to 2012.[7]

Focus Standards on Outcomes, Not Technologies

Standards should be set based on desired performance outcomes (e.g., fuel efficiency, pollutant emissions) rather than mandating the use of specific technologies. For example, rather than mandate the use of a particular type of particulate filter on trucks, impose a maximum level of particulates that may be emitted by a truck per kilometer it travels.

Specifying performance outcomes rather than mandating specific technologies gives companies the maximum leeway to innovate and apply different solutions, helping to achieve the desired outcomes at least cost (and possibly using novel techniques or technologies that were not anticipated at the time the regulation was written). It also reduces the burden on policymakers to keep up with technology developments in all fields for which they have adopted standards.

There are exceptions to this principle. When a technology is not fully mature but is ready for commercialization and shows promise to achieve very good results in the future, it may be worthwhile to create a separate performance tier or carve-out for the promising technology. This may provide it space in a crowded market, allowing it to achieve economies of scale and to develop to the point where a separate tier is no longer necessary. A good example is a renewable portfolio standard, which is also discussed in detail in Chapter 4, "Renewable Portfolio Standards and Feed-In Tariffs."

A typical renewable portfolio standard is technology-finding: It specifies a fraction of electricity that must come from clean sources (and sometimes efficiency), and utilities determine the most cost-effective way to meet the standard. For instance, a utility may choose to build some wind turbines, build some solar panels, and start a demand-side efficiency program. However, some renewable portfolio standards contain a carve-out for less-developed technology, such as offshore wind, in the form of a minimum percentage of energy that must come from that source. Even though it is not technology-finding, a carve-out may be justified in the short term, while offshore technology develops and prices come down. In time, a carve-out for offshore wind should become unnecessary.

Prevent Gaming via Simplicity and Avoiding Loopholes

While some manufacturers will voluntarily create cutting-edge products that significantly exceed performance standards, others will do as little as possible while still complying with the law. This can lead them to seek out loopholes, ways of complying with the letter of the law while undermining its purpose.

If a standard is complex and intricate, and if it makes many distinctions between different types of equipment based on various design features or use cases, this leaves many openings where companies can potentially game the law. To insulate a standard against gaming and loopholes, it is helpful to write the standard to maximize simplicity and clarity and to state in broad terms the targets that must be achieved rather than making exceptions or different rules for equipment with different features.

For example, in the United States, the Environmental Protection Agency defined different vehicle fuel economy standards for "cars" and "light trucks." Light-duty vehicles could be classified as light trucks based on minor design features (such as a vehicle that is "available with special features enabling off-street or off-highway operation and use").[8] Manufacturers were able to exploit the two different standards by making the minimum number of design changes necessary to allow them to classify many of their cars as light trucks, even though these vehicles were marketed and usually used for on-road, personal transportation. This led to a boom in SUV sales (classified as trucks) in the 1990s, until the light truck fuel economy standard started increasing in 2005.[9]

Another notorious loophole is found in sloppily designed testing standards. The Volkswagen diesel emission testing procedure was specifically designed to reduce pollution when a car was being tested, but not otherwise. Better testing, emulating real-world conditions, probably would have prevented this problem.

Economic Signals

Create a Long-Term Goal and Provide Business Certainty

Economic signals achieve energy savings through two mechanisms. First, they affect consumer behavior. For example, consumers might choose to drive less or to turn down their air conditioning in response to a fuel tax. This happens quickly after the policy comes into effect, if consumers are informed about the price changes and respond in an economically rational way.

The second mechanism involves influencing the choices consumers make when buying new products or the decisions companies make when building new equipment and, in turn, the equipment that manufacturers choose to produce and offer for sale. A carbon tax will increase demand for fuel-efficient equipment, and manufacturers will seek to respond to that demand.

Just as when they are trying to meet performance standards (discussed earlier), businesses will take action to improve their product offerings in response to an economic signal. They need time to invest in R&D activities, to change their manufacturing processes, to alter their supply chain, and to update their marketing materials. Firms need years to fully respond to a price signal that they learn about today.

Even more than with performance standards, there is often uncertainty about the permanence and magnitude of price signals. Tax rates can change dramatically based on the political preferences and economic viewpoint of the government at a given time, and subsidies are ideally structured to phase out as a technology achieves maturity. This uncertainty can interfere with businesses' plans and reduce the effectiveness of price signals. For example, suppose a tax on fossil fuels is authorized only for a few years, with the possibility of renewal. A business may decide that it is not prudent to make large investments in R&D and retooling factories if the tax might be gone by the time any newly developed products can reach the market. Even if the tax is renewed, and renewed again, the continued uncertainty surrounding each renewal will dampen businesses' responses to the tax, so it achieves less efficiency improvement than would have been caused by the same tax had it been authorized with sufficient long-term certainty in the first place. The same can be true of subsidies. The U.S. production tax credit, discussed in Chapter 5, is one example of how failing to provide long-term certainty can lead to poor investment decisions.

Similarly, a business with limited knowledge of future subsidy rates must hedge when making investments. If a feed-in tariff for wind power is extended for only a year at a time, a business will not be able to rely on it when deciding whether to commit to the construction of a large wind farm, which would take more than a year to site, permit, and build.

Generally, subsidies should be designed to phase out over time, while taxes should be designed to increase over time. When possible, the endpoint or goal of an economic incentive should be selected and explicitly specified. For example, the goal of a subsidy for wind power may be to help wind power scale to the point where it can compete economically without a subsidy. The long-term goal of a carbon tax may be to fully price in the social cost of emissions and then to remain at that level indefinitely. If a long-term goal is publicly specified, this helps businesses understand policymakers' intentions and make plans with the benefit of having this endpoint in mind.

One example of a successful incentive phase-down is the California Solar Initiative. The California Solar Initiative provided a subsidy for residential solar installations that scaled down based on the cumulative installed capacity, with a clear future trajectory and timeline. The program is widely regarded as having been a nearly optimal subsidy policy and for having maximized economic efficiency while helping scale residential solar in California.[10]

Price In the Full Value of All Negative Externalities for Each Technology or Use a Price-Finding Mechanism
Economic incentives can be structured in two ways. Either the amount of the incentive can be specified explicitly, or the policy may stipulate a mechanism by which the incentive value is found. In economic terms, the policy assumes a known quantity *or* a known price, and it uses the market to find the other value. Each mechanism is appropriate in specific cases.

If the price of a given policy objective is known (e.g., the value of the damages caused by emissions—the externality), then a tax or subsidy can be set based on that price, allowing the market to identify the quantity of activity appropriate at that price, and to identify methods to reduce emissions per unit activity. For example, a policymaker may not be able to specify the exact quantity of CO_2 emission reductions in each industry (e.g., cement, chemicals, steel) because they are each unique and complex businesses with differing abatement opportunities. However, if the policymaker has a good estimate of the harms caused by CO_2 emissions, she can set the carbon tax at that level. Each industry will be motivated to find the most cost-effective ways to reduce its own emissions until any remaining abatement options are more expensive than the tax. From a social economics standpoint, social welfare is optimized at this point (because further emission reductions would be more costly than the benefits they provide).

The second way an economic incentive may be structured is as a price-finding mechanism. If a policymaker knows how much of something she would like to achieve (such as a specific quantity of clean energy on the grid), then a price-finding mechanism can be used to identify the lowest incentive that will achieve that outcome. For example, in a reverse auction, suppliers of a good (such as clean electricity) bid against each other for the lowest subsidy they will accept. The subsidy is set at the lowest level that would achieve a sufficient quantity of clean electricity, based on the amounts each supplier offered to produce at each price.

Eliminate Unnecessary Soft Costs
Often there are significant regulatory inefficiencies or permitting challenges that raise costs, increase timelines, or discourage investment in clean technolo-

gies. These soft costs can take many forms, such as large paperwork requirements for receiving rebates, burdensome environmental quality studies, or slow permitting processes. In many instances there are good and valid reasons for these requirements. In the case of vehicle rebates, the government needs to ensure it isn't paying rebates for any vehicle more than once, and in the case of environmental quality studies, sufficient study is needed to ensure a new project will not harm endangered species, damage people's homes, or lead to environmental damages. Nevertheless, a tradeoff occurs between regulatory burden and financial attractiveness for clean technology investments.

The government should take steps to reduce these soft costs for projects that promote decarbonization of the economy. Pre-zoning specific areas for specific types of infrastructure can be a promising approach for large clean energy projects. For example, a large area of desert, absent critical wildlife habitat, might be pre-zoned as "solar-ready," and solar projects in that area can have greatly reduced requirements for project-specific permitting or approvals. Texas's Competitive Renewable Energy Zones for wind power, discussed in more detail in Chapter 5, are a successful example.[11] Another way to lower soft costs is to standardize necessary forms and allow online submission (over the internet).

Although the methods will vary based on the policy and technology being considered, policymakers should continuously look for ways to streamline processes for clean energy technologies in order to lower their costs and drive deployment.

Reward Production, Not Investment, for Clean Energy Technologies

Economic incentives for clean energy should be based on the amount of clean energy that is generated and used, not on the amount of capacity built or money invested to purchase or install clean energy infrastructure. This ensures that the incentive is paid only when these resources are actually used.

A subsidy based on installed capacity risks contributing to three problems. First, it encourages cheaper and lower-quality equipment. An inexpensive wind turbine might have the same rated power output as a higher-quality turbine, but it might break more frequently or be unable to produce as much electricity.

Second, capacity incentives promote installations in areas that may not be ideally suited to capacity growth. For example, they might encourage the placement of wind turbines in areas with worse wind speeds or where there is not sufficient transmission capacity to deliver all of the wind power to demand centers.

Third, a capacity-based subsidy eliminates the incentive to produce as much electricity as possible, because the subsidy is paid based on the size of the installation rather than the amount of electricity it generates.

Capture 100 Percent of the Market and Go Upstream or to a Pinch Point When Possible

An economic signal becomes harder to administer and is more prone to leakage—when consumers or businesses affected by the tax purchase goods from other areas without the tax—the further downstream (closer to the final point of sale) it is administered. In many cases an economic signal also can have much greater impact if administered upstream (closer to the point where the product is produced or imported). For example, a tax on coal administered at the coal mine will make the price of coal higher and induce power plants to switch to other fuels, whereas a tax on electricity based on the amount of coal generation not only is harder to administer but also is less likely to influence power plant operators to switch to other fuels. Therefore, economic signals should be administered as far upstream as possible. Sophisticated upstream actors will then mitigate the impacts of the tax by switching to less expensive options, and only the remaining impacts of the tax will be passed down to consumers via the pricing of goods.

Ensure Economic Incentives Are Liquid

Policies that offer economic signals in the form of subsidies should ensure these incentives are liquid and do not have unnecessary transaction costs. A liquid incentive is one that is easily transferable and is akin to cash. Grants or cash payments are highly liquid, but tax credits are not. For example, in the United States, to avoid the appearance of providing subsidies, the government usually offers tax credits (a reduction of income tax charged on some types of income). Often, the entities who earn the tax credits don't have enough qualifying income to fully use the tax credits. Therefore, in order to take advantage of the credits, they are forced to partner with tax equity investors, entities that have sufficient qualifying tax liabilities. These entities are generally investment banks and other large financial institutions. Having to partner with one of a limited set of tax equity partners means less of the incentive goes directly to the clean energy developers while raising project costs (finding and negotiating a contract for tax credits is expensive). It would be more efficient to just provide cash payments or grants to the developers.

Ensuring subsidies are liquid and usable by the intended recipient helps to reduce the risk and complexity faced by clean energy projects and ensures government monies are used to subsidize projects most efficiently.

Support for R&D

Government support plays an instrumental role in energy R&D progress. This support may take three forms. The government may conduct research itself, in national laboratories or similar facilities. The government may fund research undertaken by others, primarily at universities and private companies. Lastly, the government may enact policies creating a favorable environment that facilitates private companies' decisions to conduct R&D and makes their efforts more productive. The design principles for R&D are introduced here and discussed in detail in Chapter 14, "Research and Development Policies."

Create Long-Term Commitments for Research Success
Performing research and developing new technology are lengthy processes, and several years may elapse after the start of R&D for any specific technology. Support for R&D must be robust and continuous over the long term to encourage companies to invest in the personnel and equipment needed to drive innovation.

Use Peer Review to Help Set Research Priorities
To help guide funding priorities, the government should involve the private sector, which can bring crucial expertise regarding the technologies, markets, scalability, and technical challenges associated with early-stage technologies. Bringing this experience to bear on funding decisions can help ensure that government R&D dollars are spent wisely.

Use Stage-Gating to Shut Down Underperforming Projects
Because R&D is inherently risky and can involve a lot of time and resources, it is useful for firms or governments funding R&D to review projects periodically and ensure they continue to be worth investing in. Stage-gating is the process of establishing certain milestones that projects must hit before they are given

additional funding. Projects that fail to meet critical milestones are defunded, allowing resources to be spent elsewhere on more promising research.

Concentrate R&D by Type or Subject to Build Critical Mass

An efficient way for the government to fund and support R&D is to concentrate funding on a specific topic in more focused, granular institutions, possibly co-located with one another. This allows researchers working on similar technologies to share information and work together while avoiding the inefficiencies that can arise from spreading funding for similar research across many different institutions. Done well, this approach also helps establish "innovation hubs," such as Silicon Valley in California, which has become the premier global location for software and information technology innovation.

Make High-Quality Public Sector Facilities and Expertise Available to Private Firms

Many countries have already invested in expensive, high-tech equipment for R&D purposes, often owned by the government. One way to improve government and private sector R&D is to allow businesses to partner with government-owned labs to conduct research together. Partnering with private companies can help overcome some of the cost barriers that prevent companies from buying their own equipment and conducting advanced R&D.

Protect Intellectual Property without Stymieing Innovation

A strong intellectual property (IP) foundation is critical for encouraging R&D investment by private firms. If patents are not protected, then any firm can make use of research results in its own products, reducing or eliminating the incentive for firms to engage in R&D in the first place. Policymakers should ensure that patent systems are sufficiently strong and enforced at a level that encourages innovation while also taking care not to provide overly broad IP protections that encourage unnecessary litigation and stifle innovation.

Ensure Companies Have Access to High-Level STEM Talent

A highly educated labor force is a key element for successful R&D. Ideally, companies and government-owned research facilities will have a large pool of researchers to draw on with strong backgrounds in science, technology, engineering, and mathematics (STEM). To attract this talent, policymakers can establish top-quality education programs and ensure immigration laws allow companies to hire STEM talent from other countries.

Conclusion

Policymakers have many options available when it comes to climate and energy policy, but these options generally fall into four categories: performance standards, economic signals, support for R&D, and supporting policies. Performance standards set minimum performance requirements and can push more efficient and cleaner technology into the marketplace. They are particularly well suited for instances where there are significant market barriers or information is hard to come by. Economic signals, which either subsidize products and outcomes or tax inputs or emissions, can encourage adoption of more efficient technology and less polluting behavior. They are particularly effective for industries that are highly price sensitive and where there are significant substitutes available. Together, performance standards and economic signals reinforce one another to drive companies to innovate and produce better technology that makes its way into the marketplace and into the cars, factories, and power plants that make up the economy. Support for R&D can lower the costs of performance standards and economic signals while making new technologies available. Supporting policies, which vary widely, are important as well and can increase information access and push new, more efficient technologies into use.

Performance standards, economic signals, and support for R&D are most effective when designed in accordance with a set of broadly applicable design principles. These straightforward principles can help separate good policy from bad while minimizing costs.

In this chapter we have walked through the four major types of climate and energy policies and their critical design principles. With this framework in mind, we turn next to understanding how policymakers can choose from the hundreds of policy options available to them to create a strong policy portfolio that leverages the relationships between each type of policy to drive down greenhouse gas emissions.

CHAPTER THREE

How to Prioritize Policies for Emission Reduction

In Chapter 1, we evaluated the key sources of global greenhouse gas emissions and the imperative to reduce emissions to achieve a reasonable climate future with less than two degrees of warming. To hit this target, we need to reduce *cumulative* global emissions by about 50 percent relative to business-as-usual through 2050. Of course, the annual emissions curve will look different from this in 2050, with emission reductions greater than 50 percent of the business-as-usual values.

Policymakers have many options to meet this goal. In Chapter 2, we outlined the types of policies available—performance standards, economic signals, support for R&D, and supporting policies—and the critical design principles central to each. No single type of policy can tackle climate change alone. Rather, each of the policy types reinforces the others, and a strong portfolio consisting of multiple policies is the most cost-effective way to capture real-world emission reductions.

With the goals and policy types in mind, we can turn to the next question: Which policies can effectively work together in a portfolio to drive down greenhouse gas emissions? This chapter lays out a framework for identifying those policies and provides insight into how to prioritize policies for reducing emissions.

Step 1. Take a Quantitative Look at the Economy

Critical: Assess the Structure of the Economy and Sources of Emissions
Ultimately, reducing emissions comes down to decarbonizing the large energy-consuming products and processes in the physical economy: cars, buses, and other vehicles; buildings and appliances; power plants; and factories.[1]

Therefore, a process to identify and prioritize the policies that will make the biggest difference for reducing emissions in a particular country must start with an evaluation of those energy-consuming products and processes in that country. How many cars are there? How efficient are they? How do we expect the number of cars and their efficiency to change over time? And so on.

This critical first step will highlight the main sources of emissions in the particular country (much like the global assessment in Chapter 1) and the trajectory of future emissions, which can help highlight priority areas of focus. For example, one analysis of Indonesia's energy sector predicts a nearly tenfold increase in electricity demand by 2050, meaning that electricity sector policies designed to prevent coal generation from filling all or most of this demand would probably reduce emissions significantly by 2050.

Recommended: Evaluate Emission Impacts of Different Options

Energy and technology assessments need not be overly complex. A high-level understanding of the sources of energy demand and emissions can serve as a starting point for prioritizing policies. That being said, with more information, policymakers can make better policy decisions, better estimate the impacts of policies on future energy use and emissions, and thereby understand whether they are on track to achieve climate goals.

Using information from energy and technology assessments, policymakers can next move to quantitative evaluations of technology and policy abatement potential. This is an important step for robust assessment and prioritization, but in resource-constrained regions, it should be possible to sort out priorities based on the overall assessment of the economy and emission sources outlined earlier.

Tools for Evaluating Emission Impacts: The Marginal Abatement Cost Curve

A common method for evaluating greenhouse gas emission abatement potential is a technology assessment. Technology assessments evaluate the potential for specific improvements in equipment to produce greenhouse gas reductions. For example, a study of the power sector in a particular country might show that replacing coal power plants with hydropower, solar, and wind plants can reduce emissions by 200 million metric tons (MMT) of CO_2e. Technology assessments are helpful for identifying the potential greenhouse gas reductions from specific changes to equipment such as vehicles, power plants, or motors in factories.

In many instances, assessments of technology potential are paired with estimated changes in capital, operating, maintenance, and fuel costs. The cost estimates can then be used to determine the relative cost-effectiveness of different technological options. For instance, the same example study on the power sector might find that replacing the coal plants with hydropower, solar, and wind plants raises capital costs by $100 million but lowers operating costs over the lifetime of the power plants (from savings on coal as a fuel) by $200 million, for a total of $100 million in savings. In aggregate, switching to these plants would save $1 for every ton of CO_2e reduced. Similar assessments can be done for a large range of technology options and compared with each other on a marginal abatement cost curve, as in Figure 3-1, where the potential abatement guides the size of any option on the x-axis, and the cost-effectiveness of any option determines the size of any option on the y-axis.

Policymakers can use marginal abatement cost curves to evaluate the abatement potential and cost-effectiveness (in dollars per ton of greenhouse gas reduced) of specific technological improvements over a certain timeframe. The McKinsey cost curves are an example of this type of analysis.[2] Marginal abatement cost curves are a highly useful tool for providing policymakers with insight into which technologies can most reduce emissions and the relative costs of different technology options. For example, McKinsey finds that globally, improving the efficiency in motor systems in the chemical industry could abate about 250 MMT of CO_2e in 2030 and generate savings of roughly €60 per ton of CO_2e avoided.[3]

Although marginal abatement cost curves are a useful tool for policymakers, they lack critical detail on *how* policymakers can achieve the technology improvements they specify. In other words, they are technology oriented; as is common in climate and energy modeling, these curves offer a technology-by-technology look at abatement potential and cost. Although it is valuable to understand the technological potential for emission reductions from different technologies, this does not answer the critical "what to do on Monday morning" question facing policymakers: Which policies can they use to most cost-effectively cause these emission-reducing technologies to be deployed?

Tools for Evaluating Emissions Impacts: The Policy Cost Curve
The best option for policymakers to use to sort out what specific actions they should prioritize is to focus on *policies*, rather than *technologies*, using a policy cost curve. Few tools exist that take this perspective. The Energy Policy

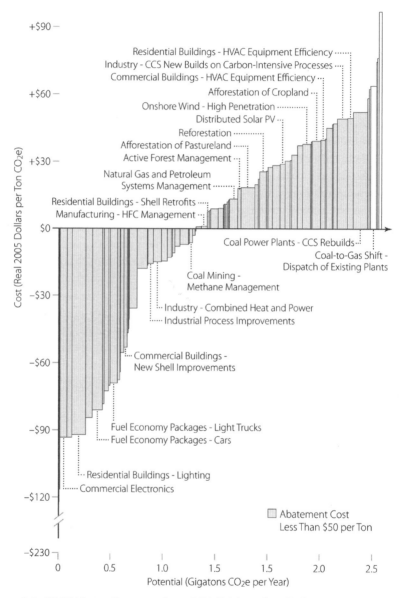

Figure 3-1. U.S. Mid-Range Abatement Curve, 2030. (Exhibit 11 from "Reducing U.S. greenhouse gas emissions: How much at what cost?," December 2007, McKinsey & Company, www .mckinsey.com. Copyright © 2018 McKinsey & Company. All rights reserved. Reprinted by permission.)

Simulator (EPS), a free, open-source, highly vetted computer model developed by the team at Energy Innovation, LLC is one such tool. The EPS is a significant advance in the field of abatement cost curves because it allows policymakers to evaluate the abatement potential and cost-effectiveness of policies rather than technologies. Although the EPS is just one such tool for evaluating energy and climate policies, we rely on it extensively throughout this book. The EPS is discussed in further detail in Appendix I.

The EPS is a system dynamics computer model that estimates the effects of various policies[4] on emissions, financial metrics, electricity system structure, and other outputs. The EPS model is designed to represent different countries by incorporating input data specific to the country in question. Appendix I discusses the purpose, structure, and function of the EPS. More detail on the technical workings of the EPS is available in the EPS online documentation.[5]

The EPS was developed by Energy Innovation, LLC with help from the Massachusetts Institute of Technology and Stanford University. The model has been peer reviewed by people associated with Argonne National Laboratory, the National Renewable Energy Laboratory, Lawrence Berkeley National Laboratory, Stanford University, China's National Center for Climate Change Strategy and International Cooperation, China's Energy Research Institute, and Climate Interactive.

The EPS aims to help policymakers evaluate a wide array of climate-related policies. The tool allows users to explore unlimited policy combinations and to adjust policy levers to any setting, allowing them to create their own policy scenarios. It simulates the years 2017–2050, using annual time steps, and offers hundreds of outputs. Some of the most important are emissions of 12 different pollutants; cash flow (costs and savings) for government, industry, and consumers;[6] capacity and generation of electricity by different types of power plants; land use changes and associated emissions or sequestration; and premature deaths avoided by reductions in particulate emissions. These outputs could help policymakers anticipate long-term impacts and costs of implementing new policies. Many of the policies included in the EPS have not yet been explored in many countries, helping to present new options to policymakers. The tool not only may inform a roadmap for policymakers to implement climate goals (e.g., from the Paris Agreement) but also may show how policymakers could set new goals and increase their countries' ambitions.

Policy cost curves require the same types of input needed to generate marginal abatement cost curves, but they include a consideration of the ability of policy to realize these technological changes on the ground. For example, in the power sector example earlier, the marginal abatement cost curve doesn't specify which policies would be most cost-effective to cause the hydropower, wind, and solar plants to displace coal, just that doing so could abate 200 MMT. By contrast, a policy cost curve would use a specific policy (or a set of policies), such as a renewable portfolio standard of 50 percent, to achieve the same emission goal in the power sector. The EPS might even find that a 50 percent renewable portfolio standard would cause a different set of renewable technologies to be built, to achieve the same emission reduction at a lower cost. The advantage of this approach is that it helps inform policymakers about the specific policy instruments they can use and what stringency is necessary for each policy to achieve the targeted emission reductions. Much like marginal abatement cost curves, policy cost curves also compare the cost-effectiveness and abatement potential of policies with one another, providing insight into which sets of policies can deliver the largest and cheapest emission reductions.

Figure 3-2 is a policy abatement cost curve for the United States generated using the EPS. It shows a cost curve for various policies, each set to a high level of stringency. Each box represents a particular policy at a particular setting—with width indicating average CO_2e emission abatement per year and height representing average costs or savings per ton of CO_2e abated through 2050.[7] Boxes below the *x*-axis represent policies calculated to save money on average through 2050, and boxes above the *x*-axis cost money.

At the far left of the curve are the cost-saving policies, which are dominated by performance standards. Performance standards often result in cost savings because in many instances the savings opportunity already exists, but other market barriers prevent action to realize those savings, like the split-incentive barrier in the building sector (discussed in Chapter 2). A carbon price is unlikely to capture these efficiency options, given that they are cost saving even without additional policy yet are still available. In these instances, a performance standard will unlock savings, where an economic signal may not be enough to overcome market barriers.

The reduction potential and costs of economic signals are clustered at the middle of the policy abatement cost curve. Economic signals, such as carbon pricing, can capture low- and medium-cost emission savings and are particularly important and effective for industries that are highly motivated by costs with low market barriers and readily available substitutes. For example, a carbon price can reduce emissions in the power sector if electricity is dispatched

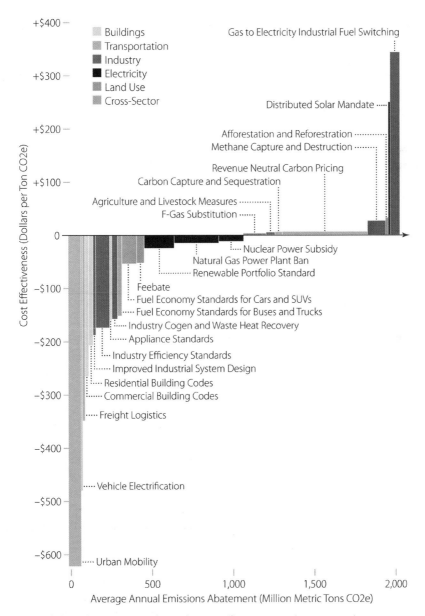

Figure 3-2. The policy cost curve shows the cost-effectiveness and emission reduction potential of different policies. (Energy Policy Simulator, Energy Innovation, January 10, 2018, https://us.energypolicy.solutions.)

based on the least-cost set of resources and the electricity mix is highly diversified, allowing lower-emitting options to displace higher-emitting ones.

Though not shown on this cost curve, R&D policies can be thought of as an overlay that decreases the costs or increases the savings of the policies included in the curve. R&D policies are intentionally excluded from policy abatement cost curves because the impact of a dollar spent on R&D on emissions is extremely uncertain. Nevertheless, R&D is critical to unlocking new technologies and to driving down the costs of existing low-carbon technologies. R&D helps to reduce the costs and increase the savings from performance standards and economic signals.

Developing marginal abatement cost curves or policy cost curves can be time consuming and expensive, and in many instances the findings may be similar to those of other countries. As a result, resource-constrained policymakers should consider looking at similar countries that have conducted quantitative policy analyses for insight. In many cases, the same set of top-level findings are consistent across countries. The patterns and policy priorities that emerge from analyzing many different countries are discussed further in Part II of this book.[8]

Marginal abatement cost curves and policy cost curves provide policymakers with insight into their available options and the abatement potential and costs of those options. But policymakers still have to prioritize the strongest and most cost-effective policies once they have this information.

Step 2. Prioritize Policies

There may be hundreds of energy and climate policies on the policy cost curve, each of which has different strengths and weaknesses and targets different outcomes. To prioritize the options, it is useful to create a hierarchy of selection criteria.

Here it is important to reiterate the need for a portfolio of policies to drive greenhouse gas emission reductions rather than any single policy. Savings from one policy can be used to pay for costs of another. Performance standards, economic signals, and R&D policies have a virtuous interactive effect as described in Chapter 2. The most effective approach will be one that leverages the benefits of different policies and policy types to reinforce others.

Policy selection criteria—in order of priority—include emission abatement potential, cost, and other considerations, outlined here.

Greenhouse Gas Abatement Potential

Above all else, policies should be prioritized based on their potential to reduce greenhouse gas emissions, because this is the primary goal of climate policy. If specific policies have dramatically higher potential abatement, policymakers should focus on these first. This will maximize emission reductions while limiting the number of policies needed to hit any emission target.

Policymakers should evaluate abatement potential in aggregate over a set number of years for two reasons. First, in terms of global warming, what ultimately matters is the cumulative amount of greenhouse gases—and their associated warming impact—in the atmosphere, not the emissions in any particular year. Second, depending on the type of policy, the emission reductions in any single year might vary widely. For example, energy efficiency standards drive growing reductions over time as existing equipment is retired and replaced by new, more efficient equipment. Looking at a snapshot in any single year is therefore likely to misrepresent the savings potential of those policies. Therefore, it is best to think in cumulative totals over time.

Notably, many of the low-cost or cost-saving reductions that can be achieved with performance standards require a long time horizon to fully take effect. More specifically and as noted earlier, performance standards often influence only new equipment, and their impact grows as the existing fleet—whether it's vehicles, air conditioners, or industrial motors—wears out, retires, and is replaced by new equipment. Because emission reductions are needed as soon as possible and performance standards take time to reach their full effectiveness but often deliver cost savings, *there is a strong case for implementing performance standards as soon as possible.*

Economic Impacts

After greenhouse gas abatement potential, the economic impact of a particular policy is the most important criterion to consider. For example, among two equally effective policies, one may have lower costs (or may drive more savings) than the other. When evaluating policies, two types of economic impacts are important to consider: direct economic and macroeconomic.

Positive or Small Negative Direct Economic Impacts

Typically, emission reductions involve some upfront capital investment that returns a stream of benefits, often involving lower energy costs (in the case of energy efficiency), lower operating costs (in the case of renewable electricity sources such as solar and wind, in which case the "fuel" is free), or increased revenue (in the case of methane capture, where additional product is available). When evaluating a policy for its economic impacts, it is necessary to consider both the upfront capital cost and the benefits that accrue over time. In many instances, energy and operating cost savings will more than pay for the upfront capital cost, even after inflation and discounting are considered. The best way to evaluate impacts is to look at the net present value of a policy, as is done in Figure 3-2, which accounts for the upfront capital cost and the stream of costs or savings that results as well as the cumulative emission abatement. A net present value provides a good approximation of the direct economic impact on individuals, businesses, and governments of complying with a given policy. Priority should be given to the policies with the lowest costs or largest savings.

Positive Macroeconomic Effects
Because climate policy affects energy choices, and energy is used in every sector of the economy, climate policies have ripple effects throughout the economy. Macroeconomic models are used to evaluate these and to provide insights into effects on jobs, wages, and overall gross domestic product (GDP), which are all important political considerations. Generally, macroeconomic modeling amplifies the direct effects of energy policies. However, some important structural shifts have been observed in response to certain climate and energy policies. Generally, the energy sector is capital intensive and produces few jobs compared with the local goods and services that make up most household spending. Therefore, even cost-neutral energy efficiency projects have positive job creation effects.[9] Other analysis has found that renewable electricity generation produces more jobs than fossil-fueled electricity generation.[10] Over the longer run, climate policy produces energy innovation, which helps make companies in the country affected by these policies better able to compete in global markets. For example, China's renewable energy push has it made a global exporter of renewable technologies and helped grow its economy.[11] Of course, if climate policy raises the prices of direct inputs for companies participating in global markets, it may be worthwhile to rebate some savings from the climate policies back to those companies, so that they can remain globally competitive. This question—how are climate policies affecting a country's companies—should be assessed and tackled on a case-by-case basis for each industry in each country.

Although macroeconomic effects may be difficult to estimate, they are worth considering when evaluating which policies to implement.

Other Considerations

Political Feasibility

It goes without saying that a key consideration of any climate policy is whether it stands a chance of being enacted. A highly abating and perfectly designed policy is not worth pursuing if there is no chance it can be implemented. To that end, policymakers should consider whether the specific policies they want to focus on have a reasonable chance not just of being passed but of being implemented, and focus on those with higher likelihood of being implemented well to drive change on the ground.

Energy Security

Many countries import a portion of their fuels from elsewhere. Being dependent on foreign sources of energy makes a country more vulnerable to supply disruptions, price shocks due to geopolitical events, and political pressure applied by trading partners. It may also skew a country's decisions about which other nations to support, or not to support, through foreign aid or military power. All told, it can reduce a country's autonomy and constrain a country's choices.

Policies that reduce the use of fuel types that are largely or partially imported help reduce these negative effects and increase the extent to which a country may pursue its own interests. Although policies that increase domestic energy production (whether clean energy or fossil energy) have the same effect, reducing dependence on fossil fuels is a better strategy, because increasing production of fossil sources would undermine emission reduction goals.

Public Health and Other Co-Benefits

Most policies that aim to reduce greenhouse gas emissions have co-benefits: positive effects for society other than mitigating climate change.[12] The most important co-benefit is usually an improvement in public health. ("Improved public health" means fewer people get sick and fewer people die prematurely.) Thermal fuels, including all fossil fuels and biomass, release harmful air pollutants when burned, such as particulate matter, nitrogen oxides, sulfur oxides, and volatile organic compounds.[13] The effects of these pollutants on human health can be estimated with epidemiological concentration–response functions and demographic data. Policymakers should take into account the number of human lives or, better, the number of quality-adjusted life years, that will be saved by implementing emission reduction policies.

Certain emission reduction policies have other co-benefits. For example, urban mobility policies that reduce the number of cars on the roads (by encouraging mode shifts to biking, walking, and public transit) reduce lost time and productivity due to sitting in traffic. These policies also help counter the sedentary lifestyles that are main contributors to the most damaging diseases in advanced economies.[14] Policymakers should think about the full spectrum of benefits offered by each policy when assembling a package of policies to drive down emissions.

Social Equity

The main purpose of greenhouse gas emission reduction policies is to avert tremendous harms from climate change that otherwise would damage human society. Low-income people around the world are particularly affected, because they do not have the resources to easily adapt to climate shifts. These impacts on vulnerable people dwarf any regressive economic effects of the emission abatement policies themselves. Accordingly, even an emission abatement policy that has a regressive economic effect is, on the whole, probably beneficial to lower-income people and justified. Nonetheless, it is ideal to structure emission reduction policies to minimize the burden they impose on low-income residents, so costs can be borne more equitably by society.

Some policies, including many performance standards, do not impose net costs over the life of the assets they target: Fuel savings more than make up for the initial difference in capital cost between a more efficient and a less-efficient model. In these cases, it may be helpful to soften the initial capital outlays via a rebate program, zero-interest financing, or similar mechanism.

Other policies, such as carbon pricing, do impose net monetary costs. Even if the cost is imposed upstream, for example at a petroleum refinery or power plant, a large portion of the cost will be passed on to consumers. Low-income consumers spend a greater fraction of their income on energy services, such as transportation, heating, cooling, lighting, and cooking. Therefore, just like sales taxes and value-added taxes, carbon pricing tends to place a greater burden on lower-income people.

The very purpose of carbon pricing is to make the financial cost of fossil fuel use better reflect its true costs to society. Therefore, it is appropriate that all people, of all income levels, should pay for the full value of these harms. Instead of providing exceptions to carbon pricing or choosing a pricing that is too low to reflect the true costs of fossil fuel combustion, policymakers can use the revenues in a way that helps offset the negative impacts of carbon pricing on low-income people. This is discussed in more detail in Chapter 13 ("Carbon Pricing").

Step 3. Put Together a Smart Portfolio of Policies

Although there are dozens of policies for policymakers to choose from, decades of experience designing and implementing policies in countries, regions, and cities around the globe has revealed that a small handful of well-chosen and well-designed policies can deliver deep decarbonization.

A pattern emerges in policy modeling: Different regions typically yield similar findings. There is of course variation in relative effectiveness of different policies based on a country's economic structure and emission sources, but the same small group of policies (implemented stringently and designed well) consistently emerge as being capable of delivering huge reductions in greenhouse gas emissions cost-effectively. So which policies are they?

Using the EPS and regional greenhouse gas emissions published by Pacific Northwest National Laboratory and the Joint Global Change Research Institute at the University of Maryland, we evaluated the global potential for climate and energy policies to lower emissions between now and 2050 and give the world at least a 50/50 chance of staying under two degrees by the end of the century.

Our analysis confirms that a small set of strong climate policies can put us on a promising emission trajectory across each of the major energy sectors. Figure 3-3 shows the emission reductions achievable from each sector.

Figure 3-4 shows the relative contribution of each top policy to meeting the global greenhouse gas emission abatement necessary to meet the two-degree target. These top policies include renewable portfolio standards and feed-in tariffs; complementary power sector policies, such as utility business model reform; vehicle performance standards; vehicle feebates; electric vehicle promoting policies; urban mobility policies, such as parking restrictions and increased funding for alternative transit modes; building codes and appliance standards; industrial energy efficiency standards; industrial process emission policies; carbon pricing; and R&D policies. This analysis is discussed in depth in Appendix II.

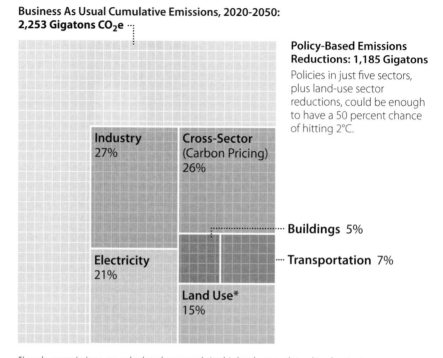

Business As Usual Cumulative Emissions, 2020-2050: 2,253 Gigatons CO₂e

Policy-Based Emissions Reductions: 1,185 Gigatons

Policies in just five sectors, plus land-use sector reductions, could be enough to have a 50 percent chance of hitting 2°C.

Industry 27%

Cross-Sector (Carbon Pricing) 26%

Buildings 5%

Electricity 21%

Transportation 7%

Land Use* 15%

*Land-use emissions are calculated separately in this book, as explained in chapter two.

Figure 3-3. Sectoral contributions to meeting the 2°C global warming target. (Analysis done using data with permission from the International Institute for Applied Systems Analysis [IIASA]. Data source: Tavoni et al., 2013. Data downloaded from the LIMITS Scenario database hosted at IIASA, https://tntcat.iiasa.ac.at/LIMITSPUBLICDB/dsd?Action=htmlpage&page=about.)

It's worth noting that the policy reduction potential in this analysis is *incremental* to existing policy. In other words, the potential of specific policies to deliver emission reductions does not account for the effect of policies already enacted. Rather, it looks at the potential for strengthening existing policies, in addition to adding new policies where they do not already exist. For example, China, Europe, and the United States already have strong fuel economy standards in place through the 2020s. Although we estimate the impact of strengthening these standards, our analysis does not attribute future emission reductions from existing standards. Everything is incremental.

Where designed and implemented well, each of the policies in Figure 3-4 has a demonstrated track record of success, driving significant emission

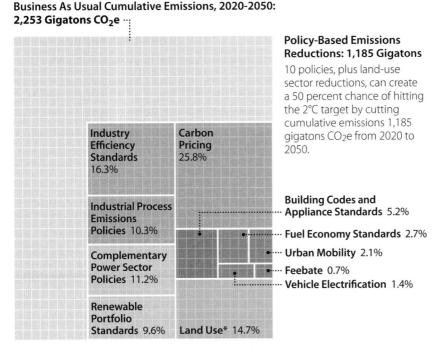

Business As Usual Cumulative Emissions, 2020-2050: 2,253 Gigatons CO$_2$e

Policy-Based Emissions Reductions: 1,185 Gigatons

10 policies, plus land-use sector reductions, can create a 50 percent chance of hitting the 2°C target by cutting cumulative emissions 1,185 gigatons CO$_2$e from 2020 to 2050.

Industry Efficiency Standards 16.3%

Carbon Pricing 25.8%

Industrial Process Emissions Policies 10.3%

Complementary Power Sector Policies 11.2%

Renewable Portfolio Standards 9.6%

Land Use* 14.7%

Building Codes and Appliance Standards 5.2%

Fuel Economy Standards 2.7%

Urban Mobility 2.1%

Feebate 0.7%

Vehicle Electrification 1.4%

*Land-use emissions are calculated separately in this book, as explained in chapter two.

Figure 3-4. Policy contributions to meeting the 2°C global warming target. (Analysis done using data with permission from the International Institute for Applied Systems Analysis [IIASA]. Data source: Tavoni et al., 2013. Data downloaded from the LIMITS Scenario database hosted at IIASA, https://tntcat.iiasa.ac.at/LIMITSPUBLICDB/dsd?Action=htmlpage&page=about.)

reductions. Consider, for example, the success of renewable portfolio standards in driving down power sector emissions in California (discussed in Chapter 4), the huge efficiency improvements achieved by Japan's Top Runner program for passenger vehicles (discussed in Chapter 6), and the vast improvement in industrial energy use from China's Top 10,000 Industries project (discussed in Chapter 11).

These policies complement each other to minimize costs of carbon abatement as well. For example, energy savings from renewable portfolio standards, building codes, and vehicle standards reduce the total cost of carbon pricing.

Part II of this book focuses on each of these policies, with detailed information on when they are most suitable and how to design them well. Each chapter also includes a set of case studies exploring good and bad implementations, demonstrating why the design principles discussed in each section are so critical to the success of each policy.

Every region is unique in its emissions and energy composition, its political mechanisms, and its ability to implement policy. In some regions, one policy might make much more sense than another. Yet time and time again, the policies discussed in this book are demonstrated to be extremely effective.

Conclusion

Strong climate and energy policies reinforce one another and drive down the costs of emission abatement. Performance standards, economic signals, R&D policies, and supporting policies should all be part of a strong policy portfolio. Quantitative assessment of a country's emission sources is a must, and further work can help identify the abatement potential, costs, and other impacts of policy options.

Policymakers should then prioritize their efforts on policies with the highest potential abatement and lowest costs. Policies with large potential abatement and long lead times that deliver economic savings should be prioritized first. Policymakers should round out a policy portfolio with additional sector-specific performance standards, strong economic signals (principally carbon pricing), and R&D and supporting policies that help lower the costs of abatement and provide additional compliance options. Other considerations, such as political feasibility, are important as well.

It is no longer a mystery which policies can most effectively reduce emissions. Decades of experience and new advances in modeling have revealed that a small set of policies, designed and implemented well, can deliver deep carbon reductions cost-effectively.

In Part II of this book, we will take a closer look at each of these policies, explore case studies where they have worked well and where they haven't, and discuss the key design principles that differentiate successful from unsuccessful policies.

The Top Policies for Greenhouse Gas Abatement

In Part I, we examined the sources of greenhouse gas emissions and why focusing efforts on industrial processes and energy use in the top 20 countries is the best approach for achieving a low-carbon future. We also discussed the different policy types—performance standards, economic signals, support for R&D, and supporting policies—and how they reinforce each other. This led to another important point: There is no silver bullet policy for solving climate change. Rather, a broad portfolio of policies is most effective for driving down emissions. Finally, we outlined ways for policymakers to study the top policies available to them for reducing emissions and how to choose between those policies. But more broadly, we showed how decades of experience, across many countries and backed by new modeling capabilities, suggest that just a small handful of policies can deliver the deep decarbonization necessary to meet the goal of staying below two degrees of warming.

Part II explores these policies in depth. Each chapter includes information on how a policy works, when to use the policy, the policy design principles most applicable and how they can be implemented, and case studies of good and bad implementation of the policy. These chapters can be used to learn more about each of the policies, their potential contribution to global emission abatement, and how to design them from a high level. For more detailed information, policymakers should look to the organizations working on specific policies. Many of the following chapters reference the work of these organizations, which can be important resources. The following chapters provide good high-level information and outline the essential elements of the top policies for tackling climate change.

THE POWER SECTOR

The power sector is responsible for 25 percent of annual global greenhouse gas emissions today, with emissions of about 12 billion tons of CO_2. Emissions are expected to grow to nearly 18.9 billion tons by 2050, comprising roughly 30 percent of annual greenhouse gas emissions in 2050. Without additional policies, the power sector will be responsible for 28 percent of cumulative emissions through 2050.[1]

The growth in emissions is caused largely by growing amounts of coal and natural gas used for power generation. For example, the U.S. Energy Information Administration projects that global electricity generation from coal will grow from 8.1 thousand terawatt-hours in 2010 to 11.1 thousand terawatt-hours in 2050, while global electricity generation from natural gas will grow from 4.6 thousand terawatt-hours in 2010 to 11.1 thousand terawatt-hours in 2050.[2]

Reducing emissions from the power sector involves using lower- or zero-carbon technologies to produce power and reduce the demand for electricity. The best policies for increasing the share of carbon-free power generation are renewable portfolio standards and feed-in tariffs, discussed in Chapter 4. Complementary power sector policies that encourage utilities to pursue cleaner options and to reduce the demand for electricity are also important and are covered in Chapter 5. Other policies that seek to reduce demand by improving the efficiency of energy-consuming products (e.g., appliances) are tackled in other sections within the appropriate sector.

The power sector has an important role to play in helping decarbonize the economy. Together, the policies discussed in this section can contribute at least 21 percent of the reductions needed to meet the two-degree target (Figure S-1).

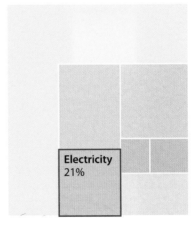

Figure S-1. Potential emission reductions from the electricity sector. (Analysis done using data with permission from the International Institute for Applied Systems Analysis [IIASA]. Data source: Tavoni et al., 2013. Data downloaded from the LIMITS Scenario database hosted at IIASA, https://tntcat.iiasa.ac.at/LIMITSPUBLICDB /dsd?Action=htmlpage&page=about.)

Renewable Portfolio Standards and Feed-In Tariffs

The electricity sector is a key part of the global greenhouse gas emissions picture; roughly 30 percent of such emissions come from burning fuels to generate electricity and heat.[1] Fortunately, the economics of this transition are rapidly improving, particularly through a transition to wind and solar. But this transition won't happen by itself: Electricity providers, primarily electric utilities, have the knowledge and capital invested in a system run primarily on fossil fuels. A strong policy signal is needed to stimulate electric utility and private sector investment in renewable energy sources as alternatives.

The two most common and successful policies for promoting renewable energy in the power sector are feed-in tariffs and renewable portfolio standards. At least 10 percent of potential reductions to achieve a reasonable chance of keeping warming under two degrees can come from these policies alone (Figure 4-1). This chapter considers both feed-in tariffs and renewable portfolio standards because their roles in promoting renewable energy are similar; each policy creates a compensation mechanism for renewable energy generation and drives renewable energy growth, although a feed-in tariff is price based, while a renewable portfolio standard is target based.

Policy Description and Goal

The goal of a renewable portfolio standard or a feed-in tariff is to stimulate the market for renewable generation, which may in turn achieve a combination of public policy goals including reducing air pollution, economic stimulus, and decarbonization. Each policy drives renewable growth, and a well-designed renewable portfolio standard or feed-in tariff stimulates the intended amount of new generation without overpaying for new resources.

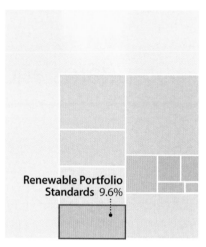

Figure 4-1. Potential emission reductions from renewable portfolio standards. (Analysis done using data with permission from the International Institute for Applied Systems Analysis [IIASA]. Data source: Tavoni et al., 2013. Data downloaded from the LIMITS Scenario database hosted at IIASA, https://tntcat.iiasa.ac.at/LIMITSPUBLICDB/dsd?Action=htmlpage&page=about.)

An important consideration for both renewable portfolio standards and feed-in tariffs is determining which resources qualify under the policy. The efficacy of either policy in driving down carbon reductions (typically a primary or secondary goal of the policy) depends to a large degree on which resources are included. Policymakers should carefully consider which technologies to include in the policy and ensure that the selected technologies will support the goals of the policy.

Other policies support the deployment of renewables as well, such as tax credits for investment in or production of renewable energy. Although these options may be valuable in certain regions, these policies are not covered in this chapter.

Feed-In Tariffs

Feed-in tariffs are price-based renewable energy procurement mechanisms, under which policymakers set and consumers pay a guaranteed price, often at a premium, for each unit of energy produced by qualifying renewable power plants. A feed-in tariff typically includes three provisions: guaranteed grid access, consistent long-term purchase agreements, and payment based on cost, value, or competitive bidding of renewable generation technologies.[2] Feed-in tariffs are typically technology specific (e.g., wind may receive one price, and solar or biomass receives another). However, policymakers may choose, for a range of reasons, to use other mechanisms, including reverse auctions, competitive bidding, or avoided cost tests, to determine a single project-level price. These mechanisms and the rationale for their use are discussed in more detail in the "Policy Design Recommendations" part of this chapter.

Key design elements for feed-in tariffs include:

- Whether and how the tariff differentiates between different technologies, project sizes, locations, and other attributes;
- Which renewable technologies are eligible to receive the tariff;
- How the tariff accounts for changes in the value of money (i.e., inflation);
- In a restructured market where generators compete to provide electricity, whether the tariff is fixed or varies with the spot price for electricity;
- Duration of feed-in tariff payments;
- The amount of energy or power available for the tariff;
- Whether and how the tariff changes over time to reflect changes in the cost or value of technology; and
- Whether the procurement relies on an auction mechanism to set the tariff or whether the price is administratively determined.

Renewable Portfolio Standards

Renewable portfolio standards set procurement targets that typically require load-serving entities (LSEs), companies such as utilities that deliver electricity to customers,[3] to procure a fixed portion of their generation from a defined set of eligible renewable sources by a certain date (e.g., "25 percent wind and solar by 2025"). The goals are typically binding on LSEs, creating penalties for noncompliance. However, some renewable portfolio standards are voluntary.

Compliance with renewable portfolio standards is generally tracked through a credit system. The LSE must hold credits that represent the renewable attributes of a unit of energy—a "renewable energy credit." Renewable energy credits are conveyed to renewable generators based on the volume of electricity generated by a qualifying power plant (e.g., one megawatt-hour of wind might generate one renewable energy credit). A renewable portfolio standard can allow LSEs to obtain credits by purchasing them on an open market, by contracting with suppliers of renewable electricity to obtain legal ownership of the credits as they are produced, or by producing their own credits by owning and operating renewable power plants. Whether a renewable portfolio standard allows renewable energy credits to be sold separately from the electricity, called unbundled renewable energy credits, depends on the specific renewable portfolio standard language.

Renewable portfolio standards generally ascribe renewable energy credits equally to all qualifying renewable technologies. However, some renewable portfolio standards allow certain technologies, project sizes, or geographies to generate extra renewable energy credits or sometimes set aside a specific share of credits that must come from a single technology type (typically called a carve-out) in order to stimulate investment in specific technologies or regions. By using a tradable credit system, a renewable portfolio standard leverages market dynamics to reduce the price of compliance for LSEs.

Key design aspects to consider for a renewable portfolio standard include:

- What entity is required to meet the renewable portfolio standard;
- Which technologies, geographies, and vintages qualify for renewable energy credits;
- Whether there are carve-outs for specific technologies;
- Whether renewable energy credits can be banked and transferred across years;
- What target and timeline are realistic given the current state of the market;
- Whether renewable energy credits are bundled (directly linked) with generation or may be unbundled and traded without regard to the purchase of renewable generation itself;
- If LSEs are allowed to purchase renewable energy credits from other jurisdictions, how the renewable portfolio standard complies with any trading or commerce rules (e.g., interstate commerce in the United States), how renewable energy credit definitions can be coordinated across jurisdictional lines, and how to avoid double-counting renewable energy credits that are traded;
- Whether there are cost caps, called alternative compliance payment rates, and what value they should be set at;
- Whether there is a minimum plant size to generate renewable energy credits and what that size is;
- Whether the entity at the point of regulation has the option to pay a fee, called an alternative compliance payment, to satisfy its requirement, what that fee is, and the use of any fees collected; and
- Whether the renewable portfolio standard should be a set target in a specific year or, instead, a rate of improvement, thereby encouraging long-term, predictable progress.

Feed-in tariffs and renewable portfolio standards can be complementary and in some cases combined. For example, some U.S. states use a feed-in tariff to support small-scale renewable energy and a renewable portfolio standard to incent utility-scale buildouts to meet overall renewable energy goals.[4] It does not make sense to use both a feed-in tariff and a renewable portfolio standard for the same technology at the same time, however.

When to Apply These Policies

When to apply a feed-in tariff or a renewable portfolio standard depends on at least three factors:

- The maturity of the domestic renewables industry
- Technological diversity and renewable resource supply
- Experience with market-based pricing in the electricity industry

Feed-In Tariffs

By driving new generation online through a set price guaranteed for a defined period, feed-in tariffs provide transparent, consistent revenue to renewable energy providers. This lowers risk to developers who may be uncertain whether they could generate sufficient revenue to cover the cost of the project over the lifetime of the investment.

The revenue certainty that feed-in tariff payments provide can be particularly valuable for nascent industries and technologies that require significant investment to move from commercial demonstration to market maturity. For example, Germany's €150-per-megawatt-hour feed-in tariff for offshore wind in 2009[5] drove a rapid expansion of offshore wind production, which has helped drive prices down significantly. From 2009 to 2016, Germany's offshore wind capacity grew from 40 megawatts to 4,130 megawatts,[6] making Germany responsible for 29 percent of global capacity.[7] In 2017, German offshore wind developers were able to put together a long-term contract without subsidy, reflecting the success of early policies in driving down prices for a new technology that ended up being competitive in the open market.[8]

However, when prices are set administratively, feed-in tariffs can do a poor job of hitting precise targets and controlling costs. In the German example described earlier, feed-in tariffs were in place for offshore wind since 2004[9] but produced virtually no investment in offshore wind facilities until 2009. On a broader scale, Germany has not managed to tightly control the amount of payments flowing to renewable energy sources. For example, in 2004 Germany set goals of 12.5 percent renewable generation by 2012 and 20 percent

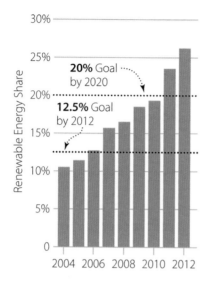

Figure 4-2. Share of renewable energy in Germany, 2004–2012. (Analysis done using data with permission from the Fraunhofer Institute, "Energy Charts," accessed December 12, 2017, https://www.energy-charts.de/power_inst.htm.)

by 2020 alongside the set prices paid via feed-in tariffs to meet those goals (Figure 4-2).[10] By the time the goals were amended in 2012, Germany had already generated 26.2 percent of its electricity from renewable sources, at significant cost to customers.[11]

As a result, Germany and other countries have implemented adjustments to their feed-in tariffs that take market development and falling costs into account.[12] These adjustments included a "reverse auction" mechanism, which can be very effective when the desired quantity of renewables in the mix is known. In a reverse auction, policymakers open a market for a fixed quantity of renewable energy, and developers bid to provide those renewables at a competitive price. The lowest bid price wins, and the funds are then rewarded as projects produce power. This avoids setting prices administratively, so it can result in much better cost efficiency.

In sum, although feed-in tariffs can be designed to support renewable investment in mature markets, they are a particularly valuable tool in nascent, less-developed renewable markets and where there is low experience with market-based pricing in the electricity industry. In these arenas, feed-in tariffs can drive investment and can help lower technology costs as increasing deployment results in economies of scale and learning by doing (where the process of building and installing new sites results in process improvements that lower costs).

Renewable Portfolio Standards

The strengths of renewable portfolio standards are twofold: policy certainty and market-based pricing. By setting standards and targets rather than prices, renewable portfolio standards create policy certainty that renewable energy goals for the power sector will be met.

Policy certainty for renewable portfolio standards depends on policy design, and it is significantly higher if three conditions are met. First, LSEs that are bound by the standard must have sufficient revenue to procure the renewable energy credits. Typically, renewable energy credits are financed by charges passed through to retail customers, which forms a consistent base of revenue. However, these charges can be fixed, based on estimated compliance costs, or they can be variable, based on actual renewable energy credit prices and the level of procurement. When renewable energy credit budgets are based on fixed charges, whether the renewable energy credits are purchased by the state (as in the case of New York) or by the LSE (as in most other regions), using a fixed charge can result in too small a budget to support the renewable portfolio standard target. When this happens, renewable energy credit purchasers will not have enough revenue to purchase the required quantity of renewable energy credits and will fail to achieve renewable portfolio standard targets.

Second, the renewable portfolio standard must have a strong enforcement signal to LSEs that fail to meet the standard. Usually this is applied through alternative compliance payments, better known as penalties, that the LSE must pay if it fails to meet renewable portfolio standard targets. These penalties must be high enough to discourage LSEs from missing renewable portfolio standard targets.

Third, the renewable portfolio standard must provide sustainable revenue to the developer, whether through long-term contracts for renewable energy or by procuring renewable energy credits over the lifetime of the renewable generator. Failure to satisfy these three conditions can result in a deficiency of renewable energy credits (i.e., not enough projects being built) that can undermine the success of the renewable portfolio standard program.

A signal benefit of a renewable portfolio standard is that it is price finding, which will control costs if the market is competitive. There are at least two kinds of market designs: a centralized spot market and auction-based renewable energy credit procurement. Spot markets create a central clearinghouse for renewable energy credits where buyers can buy renewable energy credits at least cost. Auctions publish quantities of renewable energy credits demanded by the LSE and solicit developers to compete on a one-time price to provide them, usually leading to a long-term contract. Spot markets are economically efficient in theory because renewable generators compete to provide renewable energy credits, allowing buyers to get the cheapest renewable energy credits rather than locking in prices in a contract. However, this benefit may be offset by the risk it creates for developers, which will require them to increase their bid prices.

Renewable energy credit prices can vary significantly from year to year depending on both the interaction between supply and demand and policy certainty. For example, renewable portfolio standard compliance often allows banking: Utilities may take advantage of attractive prices or available development today and use banked credits in future years, causing prices to fall. In other situations, land scarcity or lack of transmission may result in high prices. Some jurisdictions enforce renewable portfolio standards through alternative compliance payments for each megawatt-hour utilities fall short of the renewable portfolio standard. If these payments are lower than the actual cost, utilities may rationally choose to pay penalties to save their customers money. Figure 4-3 illustrates the variation in renewable energy credit prices in the northeastern United States.

Whether auctions or spot markets produce the lowest cost depends on the maturity of the renewable energy marketplace. For mature technologies that are cost-competitive with traditional power plants, renewable energy credit price fluctuations will be a modest barrier, whereas less mature technologies will probably provide energy most cost-effectively under long-term contracts.

Renewable energy credits are also generally technology neutral, which reduces compliance costs for consumers but may create barriers to entry for fledgling industries. Renewable portfolio standard auctions allow all renewable sources to compete as long as they qualify under the renewable portfolio standard. More mature, cheaper renewable technologies are therefore favored at the expense of high-potential, nascent technologies that need opportunities for deployment and higher revenue streams in order to realize significant cost declines. Not only will this crowd out potential new industries and foreclose potential supply chain or soft cost breakthroughs, it reduces the diversity of the renewable energy fleet—a crucial feature needed to reach high penetrations of renewable energy at a reasonable cost.[13]

One way to mitigate this impact is to use a carve-out, which specifies a fraction of the total renewable portfolio standard that must be achieved using a certain technology or set of technologies. For example, many renewable

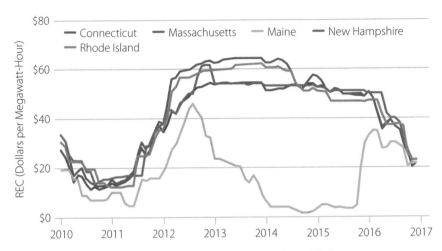

Figure 4-3. Renewable energy credit (REC) prices can vary substantially from year to year. (Graphic reproduced with permission from the Lawrence Berkeley National Laboratory.)

portfolio standards historically had carve-outs for solar photovoltaic (PV), which require a minimum fraction of the renewable portfolio standard to be met exclusively by this technology, and this helped drive the amazing price reductions in solar PV. Central station solar power plants would be a good candidate for this approach today.

Another approach is to use a tiered renewable portfolio standard, where different resources are included in different tiers. For example, tier I resources might include strictly zero-emitting technologies such as wind and solar, and tier II resources might include a broader set of renewable technologies such as biomass and waste-to-energy, with fewer credits per unit of electricity generated.

Each policy has an important role in promoting renewable energy development. Feed-in tariffs are particularly valuable in nascent, less-developed renewable markets and when there is low experience with market-based pricing in the electricity industry. A renewable portfolio standard works best for promoting renewable technologies that are not fully mature but are established. Countries considering these policies may consider a two-step or hybrid process: Create feed-in tariffs for nascent technologies and transition to a renewable portfolio standard as technologies and markets mature.

What about Nuclear Power?

At the time of this writing, renewable energy is often more cost-effective than new nuclear power. In some regions, though, new nuclear power is a cost-effective source of zero-emission energy. Additionally, many regions have already built significant amounts of nuclear capacity. For example, the United States has roughly 112 gigawatts of nuclear capacity today, and China has roughly 48 gigawatts of capacity. Worldwide, there are 359 gigawatts of nuclear capacity today.[14] Furthermore, the next generation of nuclear reactors may prove much more cost-competitive than today's reactors, opening the possibility for a significant expansion of nuclear power.

Renewable portfolio standards can be modified to include all zero-carbon generation sources, including nuclear power. When a renewable portfolio is broadened to include more technology types, it is sometimes called a clean energy standard. Under a clean energy standard, policymakers set a goal for utilities to generate a certain amount of electricity from zero-carbon electricity technologies, including nuclear power plants and renewables. As with renewable portfolio standards, clean energy standards rely on a credit system, sometimes called zero-emission credits, akin to renewable energy credits.

In some regions with competitive electricity markets and low market prices, which can happen for several reasons, existing reactors have struggled to remain profitable and are facing closure. For example, as many as two-thirds of the nuclear power plants in the United States may be operating at a loss in today's markets.[15] Existing nuclear can be one of the most cost-competitive sources of emission-free electricity in certain regions, creating a sound policy basis for paying these plants for their zero-emission attribute to ensure they remain viable.

Facing early closure of nuclear stations and the potential loss of a significant share of carbon-free generation, some U.S. states have turned to implementing clean energy standards. To date, most clean energy standards in the United States have focused on providing compensation for existing nuclear facilities. In most instances, clean energy standards have implemented a zero-emission credit exclusively for nuclear power plants, in addition to renewable energy credits, to ensure adequate revenue to keep nuclear plants in the black and avoid undermining the incentive of renewable energy credits to help drive new technologies into the market.

As renewable technologies such as solar panels and wind turbines mature over the long run, policymakers might consider evolving renewable portfolio standards to clean energy standards that focus on clean energy generation from all sources. Such a policy could help push new technologies to market while ensuring that existing carbon-free technologies are compensated for this attribute.

Policy Design Recommendations

A number of policy design principles discussed in Part I apply to both feed-in tariffs and renewable portfolio standards.

Create Long-Term Certainty to Provide Businesses with a Fair Planning Horizon

Long-term certainty is a central element of both feed-in tariffs and renewable portfolio standards. Feed-in tariffs should guarantee compensation for renewable generators for a reasonable time period (at least 10 years and up to 30) that provides investors with certainty they will receive a reasonable return on investment. Guaranteed payback periods should align with the projected lifetime of the generation technology, although they can be front-loaded to provide certainty while reducing long-term risk for the off-taker. For example, wind and solar plants typically have lifetimes of 20 years, so power purchase agreements to comply with renewable portfolio standards should be structured as 10- to 20-year fixed prices for each unit of energy. Likewise, feed-in tariffs should provide a consistent payment for each technology. Each measure lowers the financing risk, reducing the cost of capital and risk of default or poor performance.

Renewable portfolio standards must have a long time horizon in order to give LSEs enough time to comply with the standard. Procuring renewable energy credits requires building new renewable generation plants, which can take a significant amount of time. LSEs and their partners must identify and permit sites, conduct interconnection studies, and negotiate contracts over the course of a few years before construction on the plant and associated transmission infrastructure even begins. Renewable portfolio standard goals should allow this development to occur on an ambitious but reasonable timescale. A clear long-term signal will also support negotiations with renewable energy providers who need confidence they will receive consistent revenue from the renewable energy credit transaction, whether it is bilateral or spot market based. Renewable portfolio standards therefore should set targets for at least a 10-year period, with 15 or 20 years being preferable, and interim goals for intervening years.

A key aspect of long-term consistency for both a feed-in tariff and a renewable portfolio standard is ensuring that revenue is available to pay for either. One approach is to use a fixed budget to fund these programs, as described earlier, wherein funding for renewable energy purchases is generated through a fixed tariff based on an *ex ante* assessment of compliance costs. The second approach is to first obtain the amount of renewable energy needed and then collect the necessary revenue from customers. Using a fixed budget is a less effective approach because the money allocated for renewable procurement must be flexible enough to accommodate rapidly changing market conditions. In the second approach, spreading the cost over a large group of customers, who will almost certainly continue buying power from the provider, provides revenue certainty and minimal financial disruption to consumers. It also neatly ties revenue collection to system use.

Use a Price-Finding Mechanism

If countries have a renewable energy target in mind and value cost-effectiveness over other attributes, they should consider adopting a renewable portfolio standard rather than a feed-in tariff. Feed-in tariffs generally use administratively set prices to compensate renewable energy generators and therefore are not ideal price-finding mechanisms. A feed-in tariff can be modified to do a better job of adjusting to changing costs, for example, by using a reverse auction to find prices for certain technologies and then using the auction price as the tariff price.

Renewable portfolio standards using open solicitations, particularly reverse auctions, are superior price-finding mechanisms, particularly in more mature markets. In meeting a renewable portfolio standard, LSEs can "find" the price of renewable energy credits in at least two ways: reverse auctions and spot markets. The more common reverse auction sets the quantity and defines the characteristics (e.g., development deadline, location, duration) of renewable generation desired and opens up a bidding auction through which prospective developers offer a price per unit of generation over the lifetime of the asset to the LSE. Once all bids are received, the LSE chooses the developers who meet the desired characteristics and bid the lowest prices, and it negotiates long-term contracts for the renewable energy credits and associated generation with the auction price as the basis. As long as there is robust participation in the auction, or entrants are not arbitrarily barred from participating, reverse auctions are powerful, effective price-finding mechanisms.

However, auctions are cost-effective only when there is adequate participation. LSEs are often the only buyer in a given country or region, so policymakers must take care to ensure that the LSE is operating auctions in a way that encourages competition. In all cases, LSEs must publicize the auction far ahead of time, including reaching out to prospective developers operating in the region well in advance of the auction, clearly defining the bid requirements, and sharing pro forma bids with customers to facilitate the bidding process. For LSEs that own and operate renewable generators, regulators must also decide whether the LSE should be allowed to participate in auctions against competitors and, if so, how to ensure fairness around the LSE's final selection.

Spot markets for renewable energy credits find prices through the interaction between supply and demand over time. These markets require a high degree of market maturity because the value of the renewable energy credit will change over time, introducing risk for developers. Similarly, renewable power producers may bid higher renewable energy credit prices to account for this risk, which would increase renewable energy credit prices. Still, spot markets can be useful for driving down renewable portfolio standard compliance costs, particularly when they span multiple, coordinated regions, increasing supply-side competition. For example, an interstate renewable energy credit spot market would allow LSEs in one country to comply with a renewable portfolio standard by taking advantage of plentiful resources in another.[16]

Eliminate Unnecessary Soft Costs
There are at least three key aspects to reducing soft costs for renewable energy that support the success of renewable portfolio standards and feed-in tariffs: siting, transmission access, and transaction costs. Siting renewable energy, or finding a place to put it, can be a time-consuming regulatory process that adds risk and costs to development. Permitting and site contracting add significant costs and delays and may ultimately kill projects, particularly for large-scale renewable energy and the transmission to access world-class wind and solar resources. Transaction costs are higher when renewable developers face uncertainty over whether they can sell their kilowatt-hours to creditworthy counterparts over the project's useful life. Chapter 5, "Complementary Power Sector

Policies," contains a section on investment-grade policy that expands on these strategies.

Reward Production, Not Investment, for Clean Energy Technologies

Both renewable portfolio standards and feed-in tariffs generally are structured to reward production, not investment in clean energy technologies. Feed-in tariffs are structured to compensate renewable generators based on each unit of electricity generated. Similarly, a renewable energy credit is created for each unit of electricity generated by a renewable power plant.

These policies can be contrasted with capacity targets on one hand and investment incentives on the other. Capacity targets mandate the construction of a given quantity of renewable power capacity. These targets create several problems, many of which have been evident in China (discussed later in this chapter). First, although they may stimulate rapid market development, they do not necessarily guarantee the renewable generator is connected to the grid or generating power at full technical potential. Second, unless projects are owned by public utilities, capacity targets may not provide the consistent revenue streams needed to reduce the cost of renewable investment.

Likewise, investment incentives on a capacity basis can be effective stimulants for clean energy investors. However, investment incentives may not guarantee or even incentivize system performance along the lines of developer projections. For example, although investors may receive an incentive to build a large wind farm, the off-taker (the company purchasing the electricity directly from the power plant) may still bear the risk of maintenance issues, undergeneration or overgeneration, and inaccurate wind forecasting.

Ensure Economic Incentives Are Liquid

Renewable portfolio standards and feed-in tariffs must balance incentive liquidity with the need to encourage consistent performance over the lifetime of the asset. Making economic incentives available for recovery too early may reduce developer motivation to perform for the full lifetime of the system, and drawing incentives out for too long may increase developer risk and system costs.

Feed-in tariffs are sometimes structured as tax credits rather than direct payments for each unit of electricity generated. For example, in the United States wind power receives a production tax credit of $23 per megawatt-hour of electricity generation. Often, the entities that earn the tax credits don't have enough qualifying income to offset with the tax credits. Therefore, to take advantage of the credits they are forced to partner with tax equity investors, entities that have sufficient qualifying tax liabilities. These entities are generally investment banks and other large financial institutions.

There are several problems with this approach. First, the tax equity investor typically takes more than 50 percent of the value of the tax credits, so less than half of the government's subsidy ends up being used to promote the subsidized activity. Second, because of problems in financial markets, tax equity investors may be unable to maintain the required level of taxable income over the period of the contract. Third, it is a barrier to small- and mid-scale renewable projects because tax equity investors are not interested in becoming involved in small projects.[17] By using tax credits instead of grants, this program increases project costs and underutilizes taxpayer money. Tax credits and other financing instruments are discussed in more detail in Chapter 5.

Build In Continuous Improvement

Feed-in tariffs should improve over time by closely tracking the cost of installations and adjusting the administrative price accordingly. Policymakers should continue providing compensation but ramp down prices over time as confidence improves, price points become more apparent, and industries mature. Likewise, feed-in tariffs should expand to cover new technologies that are reaching commercialization but lack a market for deployment.

Continuous improvement for renewable portfolio standards can take two forms: First, renewable portfolio standard targets can be raised at regular intervals—although this should be done well before the expiration of the older targets, to keep the markets healthy. A more powerful method would be to set an annual increase in required renewable energy credits, such as 3 percent per year, rather than a plateau target. This would create a great investment signal and would make maximum use of political capital at the time the program is established.

Finally, it is important that a renewable portfolio standard not just require compliance at the end of the program year; instead, it should implement consistent interim goals to provide consistent signals to investors and avoid dropoffs if LSEs take an irregular, particularly back-loaded, procurement strategy. It may be advisable to create a banking-and-borrowing system to allow for the lumpiness of project development. Banking and borrowing allow overcompliance or undercompliance in any given year, as long as the total renewable energy credits match the total renewable energy credit requirement over a multiyear (e.g., 4-year) period.

Focus Standards on Outcomes, Not Technologies

This principle applies specifically to renewable portfolio standards. By placing the onus on LSEs to procure a portion of renewables regardless of type, renewable portfolio standards do a superior job of following a technology-finding path and reducing costs compared with feed-in tariffs. For example, California saw wind and biomass production outpace solar in the initial phase of its renewable portfolio standard from 2002 to 2012, but as solar costs have dropped dramatically, solar has overtaken wind and is now the most prevalent technology used to meet the renewable portfolio standard (Figure 4-4).

In specific instances, it may make sense to use carve-outs for specific technologies (which violates this principle), but over time as technologies mature and costs fall, carve-outs should be phased out.

Prevent Gaming via Simplicity and Avoiding Loopholes

For renewable energy credit markets to function properly and stimulate additional investment, there must be a clear tracking system for the renewable energy credits.[18] In the United States, each renewable megawatt-hour receives a unique ID number for each renewable energy credit, based on electronic data supplied by transmission operators that can verify the generation. The tracking system then "retires" renewable energy credits when they are purchased, usually by an LSE complying with the renewable portfolio standard. These robust systems prevent double counting, and allow trading across borders if tracking systems are linked and renewable portfolio standard requirements are sufficiently similar, much like cross-border carbon markets.

Renewable energy credit markets should also have clear deliverability requirements. Policymakers must decide whether renewable energy credits must be deliverable to the grid in the policy region or whether unbundled renewable energy credits (when purchasing renewable energy credits is separate from purchasing electricity) from other regions qualify under the renewable portfolio standard. For example, the lowest-cost compliance option is usually to allow renewable energy credits from all regions qualified by policymakers to participate in a renewable portfolio standard. However, this will result in some projects being developed outside the policymaker's region. Consequently, the renewable portfolio standard may have only limited economic or emission

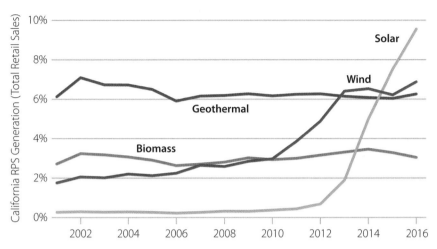

Figure 4-4. California's renewable portfolio standard (RPS) allows the lowest-cost resources to meet the target as prices evolve. ("Electricity Data Browser," U.S. Energy Information Administration, accessed December 12, 2017, https://www.eia.gov/electricity/data/browser/.)

impact in the region in which it is established. Similarly, allowing renewable energy credits from multiple regions under a single region's renewable portfolio standard requires that the participating regions have similar definitions of renewable resources. For example, Pennsylvania includes coalbed methane as a renewable energy source, but other states do not allow this resource to qualify under their renewable portfolio standards. Thus, if interregional renewable energy credits are allowed under a renewable portfolio standard program, policymakers must identify a way to align different renewable portfolio standard definitions.

Case Studies

United States (California)
The United States has no national renewable energy policy other than a feed-in tariff for wind generation that takes the form of a production tax credit of $23 per megawatt-hour and a capacity-based investment tax credit for other renewable generation technologies. Instead, the United States leaves most renewable energy policy to the states, which have significant control over their energy mix.

A total of 29 states have renewable portfolio standards that vary in structure and ambition, covering 55 percent of national generation.[19] Some have carve-outs for certain types of resources, and others leave mature technologies such as large hydroelectric generation out of the standard. Most states have adopted net metering, under which small-scale solar generation receives credit on customer bills at the full retail rate of electricity. Together, these policies have created a diverse, dynamic investment environment where renewables have thrived.

California was an early adopter of the renewable portfolio standards in 2002 and has been a leader in renewable energy ever since; today California leads the United States in renewable energy capacity and generation and has the most solar generation by percentage, at 13.8 percent in 2016,[20] due in large part to continuous improvement in the state's policy. California currently has a renewable portfolio standard of 50 percent renewable generation by 2030, excluding large hydroelectric power and small-scale customer-owned generation, which receives a feed-in tariff through net metering. California's LSEs have consistently outpaced their renewable portfolio standard requirements, taking advantage of an attractive investment environment and raising certainty of compliance. This high performance gave California legislators confidence to raise the initial renewable portfolio standard of 20 percent by 2017 (enacted 2002) to 20 percent by 2010 (enacted 2006), 33 percent by 2020 (enacted 2011), and finally 50 percent by 2030 (enacted 2015).

The California renewable portfolio standard uses a technology- and price-finding reverse auction system whereby LSEs issue requests for generator bids, and qualifying generators respond with bids for long-term contracts for both renewable energy credits and the underlying generation ("bundled" renewable energy credits) that allow LSEs to meet their generation mandates. This has kept compliance costs low; for example, from 2012 to 2014, California's renewable portfolio standard compliance yielded cost savings.[21] However, California's compliance options are limited by geographic constraints. Policies to generate at least 70 percent of the renewable energy credits in state[22] help promote economic and industry development, but they do limit the options for out-of-state, possibly more economical resources to compete and may increase compliance costs.

Additionally, California's renewable portfolio standard is technology finding, resulting in a rapid procurement of solar, because there are few other competitive options to choose from in-state. But renewable portfolio standard compliance costs have risen, even as the share of solar has grown with simultaneous cost declines. This is also creating problems for the grid: A glut of daytime solar generation on low-demand days has led to curtailment of solar. New complementary policies to diversify the resource mix, particularly greater access to out-of-state renewable resources such as wind, geothermal, and hydro, can help lower the costs of meeting the 50 percent renewable portfolio standard.

China

China is an international leader in both coal-fired and renewable energy. China has the most renewable energy capacity of any country, including 148 gigawatts of wind, 77 gigawatts of solar, and 332 gigawatts of conventional hydroelectric as of 2016.[23] China is aiming to continue this development, with a goal of surpassing 210 gigawatts of wind and 110 gigawatts of solar by 2020.

China has three complementary renewable energy policies: a feed-in tariff, renewable generation goals, and renewable capacity targets. China's feed-in tariff is a conventional design; since 2003 renewable technologies have received compensation per kilowatt-hour generated, at rates particular to each technology. To ensure the feed-in tariff is following the cost closely, Chinese state-owned grid companies issue auctions for certain technologies to find the price, and this new auction result sets a new all-in price for renewable energy from that technology. In addition, the feed-in tariff varies for different regions based on resource availability. In wind-rich regions, for example, the feed-in tariff is lower and increases where wind is scarcer.

However, despite consistent compensation for generation, curtailment of available wind and solar generation has been a persistent problem in China, because the capacity targets reward investment, not performance. For example, in 2016, 21 percent of wind power was wasted.[24] Although developers have every incentive to perform under the feed-in tariff structure, the grid managers, LSEs, and local governments have little incentive to accept the energy and pay the generators, as the overriding concern for government officials and state-owned enterprises has been meeting the capacity targets from the 13th Five-Year Plan. As a result of rewarding capacity and not generation, China's investment in new renewable energy facilities has equaled or outpaced the rest of the world since 2010, but many generators are unconnected to the grid or are routinely curtailed in favor of thermal generation, often coal.

China's system falls short of several other design principles; by relying on capacity targets and a technology-specific feed-in tariff, it fails to be technology and price finding. Recently, the Chinese government signaled that it may begin adjusting feed-in tariffs more often, improving the price-finding quality of its renewable energy policy. In addition, recognizing the curtailment issue, China has begun to experiment with a renewable portfolio standard–driven renewable energy credit trading system to allow interregional renewable trading and reduce reliance on technology-specific feed-in tariffs and capacity targets.[25]

Still, China has seen important successes, particularly in reducing soft costs and applying continuous improvement. The proactive development of east–west transmission helps reduce soft costs by providing access to wind-rich generation and transmitting it directly to load centers where it can be consumed. Furthermore, Chinese developers get access to low-interest loans from state-owned development banks, reducing the cost of capital for new projects. Continuous improvement of the capacity targets has likewise resulted in consistent progress and renewable investment that laid the foundation for tremendous success. In driving a staggering share of global demand, China has also become the dominant solar PV manufacturer, with 70 percent of global production capacity in 2012.[26]

Germany

Germany has some of the most ambitious renewable energy targets in the world, at 40–45 percent by 2025 and 55–60 percent by 2035. So far, Germany's feed-in tariff has been the principal driver for its rapid adoption of renewable energy since it began setting renewable energy goals in 2000. The feed-in tariff is set for a 20-year term, it varies by technology, and the tariff level is set at regular intervals.

Until recently, Germany has not had a price-finding feed-in tariff, and as a result the cost impacts of its rapid adoption of renewable generation have been high. Today, the renewable surcharge on customer bills averages just under 25 percent of the total electricity rate, although many industrial customers are exempt from this surcharge to remain globally competitive. To mitigate these costs and shift some risk onto maturing renewable generation industries, Germany moved from a pure feed-in tariff to a feed-in premium, which lowers the amount of compensation but allows generators to also collect revenue from wholesale energy markets. As a result, renewable generators whose costs are covered largely by the feed-in premium can compete in wholesale energy markets, driving down the price of electricity for consumers.

To become more price-finding, Germany changed its feed-in tariff to adjust downward based on renewable energy auction results, conducted annually.[27] Additionally, feed-in tariffs are front-loaded, with higher compensation for the first 5–10 years, depending on technology, and reduced each year thereafter. In these ways, Germany hopes to reduce the cost of its renewable transition while stimulating investment sufficient to meet its decarbonization goals.

Conclusion

It is clear from experience that both feed-in tariffs and renewable portfolio standards can be effective, low-cost options for accelerating the renewable electricity transition. A country with immature renewable energy markets may benefit more from a feed-in tariff, which minimizes developer risks and stimulates growth of all renewable technologies. A renewable portfolio standard may be more appropriate as the market matures, allowing the lowest-cost technology to meet a country's renewable energy goals. Each policy can be tweaked to compensate for its perceived weaknesses; in many cases, the two policies can even be used successfully together.

Decarbonizing the electricity sector is a fundamental step toward deep economy-wide decarbonization. As the carbon intensity of the electricity sector decreases, the electrification of other sectors, particularly building heating and transportation, becomes an increasingly effective decarbonization strategy. Both policies can get countries far along the path to decarbonizing the electricity sector; whether that is cost-effective depends as much on the policy designs contained in this chapter as on a country's ability to adapt its renewable energy policies to changing market conditions.

CHAPTER FIVE

Complementary Power Sector Policies

Electricity systems in many nations are facing a radical makeover that will drastically reduce greenhouse gas emissions, increase system flexibility, incorporate new technologies, and shake up existing utility business models. But international power systems are remarkably diverse, and there will be no one-size-fits-all solution for this transformation. Conversations about the best ways to lower costs, keep the lights on, and deliver a cleaner power system are often plagued by arguments over whether utilities or markets are king or whether legislators or regulators should drive system evolution. There is no "right" answer to these questions: Electricity policy is heterogeneous and will remain so.

Dramatic drops in the cost of clean energy have made it possible to deliver a very low-carbon electricity system at roughly the same cost and reliability as a dirty one.[1] But zeroing out carbon emissions from the electricity system is not as simple as setting a renewable portfolio standard at 100 percent. Even under the best possible policy design for renewable portfolio standards or feed-in tariffs, a more holistic approach is needed to ensure the transition is affordable, increases prosperity, maintains reliability, and expands service to unserved customers.

These complementary power sector policies have a large role to play in decarbonizing the economy and can contribute at least 11 percent of the reductions necessary to have a 50/50 shot at staying under 2°C of warming (Figure 5-1).

Policy Description and Goal

A more holistic plan will require energy policies beyond feed-in tariffs and renewable portfolio standards. Other changes to the power sector can accelerate renewable portfolio standards and feed-in tariffs and enable further emission reductions. The most important among these are policies promoting grid flexibility, performance-based regulation, well-functioning competitive power markets, policies facilitating orderly retirements of existing power plants, and investment-grade policy design that de-risks renewable energy projects.

These policies together can help accelerate the power sector transformation and deliver at least 11 percent of the reductions needed to meet the two-degree target.

The goal of the policies described in this chapter is to support the institutional and physical system needed for clean electricity to thrive at low cost and high reliability. To achieve this goal, policymakers can focus on five areas of activity: supporting the

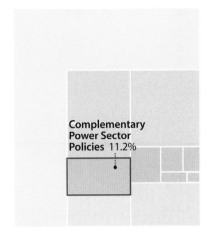

Figure 5-1. Potential emission reductions from complementary electricity sector policies. (Analysis done using data with permission from the International Institute for Applied Systems Analysis [IIASA]. Data source: Tavoni et al., 2013. Data downloaded from the LIMITS Scenario database hosted at IIASA, https://tntcat.iiasa .ac.at/LIMITSPUBLICDB/dsd?Action=htmlpage &page=about.)

many sources of flexibility on the grid, using performance-based regulation to align utility financial incentives with a clean system, designing and operating competitive power markets well, providing for orderly retirements of outdated power plants, and de-risking renewable energy development with investment-grade policy.

Grid Flexibility

A decarbonized power system will necessarily eliminate coal (at least without carbon capture and sequestration, which today remains quite expensive) and greatly reduce natural gas generation. Nuclear may remain a part of the decarbonized generation mix, but today's costs are quite high compared with those of other zero-carbon power plants. Solar and wind power are cheap options today, but their production varies with the availability of sunlight and wind, so they require a more flexible power system to realize their value as power system decarbonizers.

The electric grid has always been somewhat flexible in order to meet variable

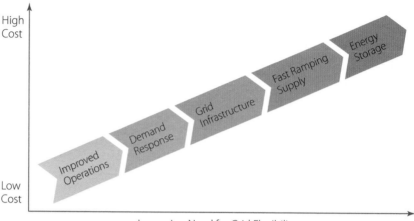

Figure 5-2. Illustrative costs of grid flexibility options. (Graphic reproduced from publicly available U.S. government data, adapted from Paul Denholm et al., "The Role of Energy Storage with Renewable Electricity Generation," National Renewable Energy Laboratory, 2010, https://www.nrel.gov/docs/fy10osti/47187.pdf.)

electricity demand in every instant. Fortunately, many options are already available to draw flexibility out of the power system. Grid flexibility can come from physical assets, such as batteries and fast-ramping natural gas plants, or it can come from improved operations, such as using advanced information technology to better co-optimize power supply and power demand.

Figure 5-2 shows options for grid flexibility in rough order of today's costs. The cheapest option is to improve grid operations to draw out latent flexibility in the existing system. This includes shortening the amount of time between each grid rebalancing and dispatch, incorporating better real-time weather forecasting into grid operations, and expanding the geographic area over which grid operators keep the system in balance if there is already transmission capacity available (to take advantage of a more diverse portfolio of resources).

Demand response is the next cheapest option. This refers to a suite of approaches to differentiate, categorize, and aggregate electricity demand in order to shift it away from times when supply is scarce to times when supply is in surplus. For example, switches and radios can turn every building into a thermal battery; by precooling or preheating buildings and water supplies, thermostats and water heaters become amazing sources of grid flexibility while delivering the same comfort and service to building occupants.[2]

After demand response comes grid infrastructure. Improved transmission and distribution infrastructure can also increase grid flexibility. Increased transmission capacity allows electricity to be transported more readily within a balancing area, meaning that more of an area's resources can be used to help balance supply and demand. Similarly, increased transmission capacity connecting balancing areas means that operators in different regions can buy and sell electricity from each other. This allows operators to draw on the resources of multiple regions to balance out variability and similarly allows operators to import electricity when local prices are high or export electricity when there is a surplus and prices are low.

Distribution system infrastructure also helps balance out supply and demand, similar to investments on the transmission system. For example, updating infrastructure to enable a two-way flow of electricity (rather than the traditional one-way flow from big power plants to customers) creates an opportunity for distributed energy such as rooftop solar or batteries located in buildings to serve more than just on-site energy needs.

Next, in terms of cost-effectiveness, is fast-ramping generation. Combined-cycle natural gas plants can provide this service, as can some hydroelectric generators, if environmental constraints are met and plants are properly compensated. And finally, storage can tackle grid flexibility needs. The cost of battery storage decreased more than 80 percent from 2011 to 2017[3] and has even begun to win out on costs in certain all-source (cross-technology) bids for flexible resources.[4]

More sources of flexibility will probably emerge as the global power system transitions. Policymakers should look for creative opportunities to improve grid operations, draw out demand response,[5] build necessary transmission and undertake important distribution system upgrades,[6] support power plants to operate flexibly when they can, and facilitate deployment of energy storage.

Performance-Based Regulation

Some aspects of electricity delivery are natural monopolies. For example, it does not make sense to build multiple sets of wires to every building. So unless the government directly owns utility infrastructure in a region, regulation of private monopoly utilities will always be a part of electricity policy. In many parts of the world, private monopoly utilities have traditionally been regulated under a "cost of service" model, which allows utilities to earn a rate of return on all capital investments and to pass through most operational expenses to customers. This structure works well when the goal is power system expansion, but it creates a bias for utility-owned, capital-intensive infrastructure.[7]

To tackle these biases and align utility financial incentives in pursuit of important societal outcomes, several regions have adopted performance-based regulation, and this regulatory structure is providing new opportunities for innovation in the utility sector. Performance-based regulation shifts some of the utility's profit incentives toward achievement of top goals.[8] For example, regulators could offer utilities cash or extra basis points on the rate of return if they meet quantitative targets for things such as CO_2 per capita, and similarly, regulators could assess penalties for failure to reach important targets. Regulators can use this kind of mechanism to shift an appropriate amount of risk onto the utility for meeting important societal goals, enabling them to use their role as market makers to drive outcomes in the public interest.

Performance-based regulation can (and should) be designed with customer affordability in mind. For example, regulators can limit a utility's revenue—called a revenue cap—in order to encourage them to behave more efficiently and keep costs low. Over time, the cap can be adjusted to account for changes in the productivity of the utility.[9] This structure contains costs and makes utilities most profitable when they serve as a platform for customers to access clean energy and demand management services. As former U.S. utility regulators Ron Binz and Ron Lehr say, performance-based regulation shifts the central question from "Did I pay the right amount for what I got?" to "Am I paying for what I want?"[10]

A well-developed area of performance ripe for measuring and compensating utilities is energy efficiency. Efficiency investments counteract traditional utility regulation because efficiency reduces sales and can avoid the need to expand capital investment by reducing wear and system peak. Because distribution utilities also handle retail sales in most places, they generally have sufficient data to identify customers who could benefit from efficiency investments based on their usage, but they would prefer not to encourage behavior changes or customer investments that reduce their profits. In the United States, it is becoming common to remedy this conflict by providing financial incentives to utilities for achieving greater efficiency via direct investment, customer education, and connecting efficiency vendors to customers. Twenty-seven U.S. state-level programs have resulted in increased spending on efficiency and cost-effective emission reductions.[11]

A financial structure that pays utilities for outcomes can set the table for innovation in a historically staid industry. Well-designed performance-based regulation aligns utility decision making to accomplish a clearly defined goal, opening up opportunities to innovate and integrate new technologies.

Well-Functioning Competitive Power Markets

Competition has moved—at varying paces in different parts of the world— into electricity generation, transmission, and demand. Many countries have introduced competitive generation, with market regulators setting rules for wholesale clearinghouses in which generators compete. Some areas have also introduced competitive transmission, wherein independent transmission companies may compete to build and operate transmission lines, taking bids and negotiating contracts to move electricity. Some have adopted retail choice, where residential and small business customers can choose their own power supplier.

As a rule, a system optimizer—often known as a regional transmission organization—is needed to facilitate competition in a region. Regional transmission organizations direct traffic on the high-voltage transmission system by dispatching power plants in a fair, reliable, and economically efficient manner. They do so by providing a neutral platform for wholesale transactions, facilitating competition between generators, transmission, and other resource providers, protecting against market power, and prioritizing the dispatch of least-cost resources.[12]

Well-managed competitive markets have the potential to lower prices, drive innovation, serve customers well, and reduce emissions. But it is tricky to design markets that cover all the near- and long-term system needs, so regulators need to act with care and sophistication to ensure markets are set up well to meet clean energy goals while balancing economic efficiency with reliability.[13] It also may be beyond the power of electricity market operators to price some externalities, particularly environmental ones. This raises the importance of promoting a technology-neutral market while using other mechanisms, such as a carbon tax or cap (discussed in Chapter 13), to value greenhouse gas mitigation.

The interface between transmission and distribution grids demands special care, if resources at the distribution scale are to provide their full value to the integrated grid. This can provide a great opportunity for low-cost, clean grid services, but capturing that opportunity presents two challenges. The first involves creating rules for these distributed resources to participate in large-scale competitive markets. The second involves operational coordination between the bulk system and distribution system, which do not currently communicate with one another on a technical and operational level.

To adapt to the changing resource mix, competitive power markets must update their rules and product definitions to expose the value of grid flexibility latent in today's power system.[14] In particular, the service of shifting supply and demand from periods of excess to shortage periods will be particularly valuable.[15] This includes flexibility on fast timescales to deal with short-term variations in wind and solar production as well as longer-term, more predictable fluctuations caused by sunset or seasonal wind patterns. Regulators can also begin to think about the long-term changes that will be required if the power system is to operate cost-effectively with a high share of low- or zero-marginal-cost resources.[16]

Orderly Retirements

As power system infrastructure turns over to become cleaner and more resilient, regulators must create a supportive environment for orderly and timely retirements of polluting power plants. In vertically integrated regions, the choices utilities and their regulators make about these retirements can have important implications for customer affordability. Analysis by the Climate Policy Initiative in the United States finds about half of the nation's coal plants are uneconomic *on a marginal cost basis* compared with the all-in capital and operational costs of a new wind facility in a nearby region.[17] But investment inertia is keeping many of these plants operating nevertheless. In regions where generation is owned by regulated utilities, allowing utilities to sell the remaining undepreciated balance of an old, uneconomical power plant to a bond-holder can enable the utility to recycle that capital into more productive, cleaner alternatives at a savings to customers.[18]

In regions with competitive generation, policymakers can support reasonable retirements by holding firm on market rules and products that maintain a level playing field, avoiding changes requested by owners of power plants that can no longer compete.[19] A well-functioning competitive power market will send the appropriate price signals to keep the system in equilibrium as old plants retire.[20]

Investment-Grade Policy: De-Risking Renewable Energy Projects

Renewable energy technologies have high initial costs but then cost very little to operate, because they do not require fuel. One consequence of this capital-intensive nature is that renewable energy is very sensitive to the cost of capital, that is, the interest rates or return rates demanded by those who lend or pay for renewable energy technology up front. For example, high interest rates can significantly increase the overall cost of a wind farm (Figure 5-3).[21,22]

Return requirements and interest rates, in turn, are driven by risk. Investors properly demand higher returns when they face higher risks. So, if smart public policy can drive down risk, it can drive down cost. And the difference can be dramatic, cutting overall costs by close to 50 percent in some cases.

Risk comes in many flavors: Technology can fail, it can be diffi-

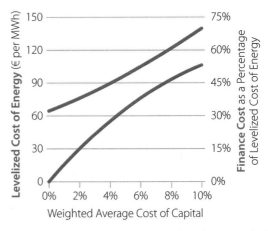

Figure 5-3. Lower financing costs can reduce the costs to build new energy technologies. (Graphic reproduced with permission of BVG Associates. Giles Hundleby, "LCOE and WACC [Weighted Average Cost of Capital]," BVG Associates [blog], 2016, https://bvgassociates.com/lcoe-weighted-average-cost-capital-wacc/.)

cult to site a wind farm, construction may be delayed because of permit problems, and the sale price of electricity might be unknown. Most of these risks can be mitigated by smart public policy, without compromising key public values. And where that is done, clean energy becomes cheaper.

The first of these risks, technology risk, is substantially reduced for solar PV and onshore wind, which are now reliable and inexpensive, and underwritten by a wide range of equity investors and project finance lenders. Other technologies are still approaching this status in many markets, including concentrated thermal solar and offshore wind. Chapter 14 ("Research and Development Policies"), Chapter 3 ("How to Prioritize Policies for Emission Reduction"), and Chapter 4 ("Renewable Portfolio Standards and Feed-In Tariffs") cover approaches to mitigate technology risk for options that are not yet competitive. This section focuses on clean energy technologies with low technology risk, which can nevertheless suffer from development and market uncertainty in the form of project development risk and price certainty risk.

Project Development Risk
Development risk comes in three principal forms: siting, permitting, and transmission access. Siting risk has many dimensions, such as landownership,

usage terms and rights, access to roads and transmission, and environmental or cultural conflicts. Each conflict raises uncertainty—sometimes a decade's worth—that can kill a project. If a developer has to spend 2 years trying to get siting questions answered, that means 2 years with no returns, 2 years of climbing expenses, 2 years of expiring tax breaks, and so forth. Good policy can drastically cut this uncertainty by pre-zoning land and engaging stakeholders early, while setting clear requirements and timeframes for permits.

Once a site has been selected and approved, dozens of other permitting requirements arise, including access to land, construction standards, fill issues, inspections, noise, traffic, visibility, dust, worker protection, and so forth. And these permits are usually required by many different federal, state, county, and city offices. The upshot can be a paper blizzard that adds years to a project. A jurisdiction with a goal of deploying clean energy can clear out this sort of costly clutter by thinking ahead, setting clear standards, and then offering rapid permits for projects that meet those standards. This is not a simple process, but it can have profound risk-reducing effects.

Large-scale solar and wind projects also need ready access to transmission lines to get their power to market. For sites without ready access, the aforementioned siting and permitting issues apply doubly to additional transmission projects. Building transmission lines that link the best, low-conflict renewable resource areas with high-demand cities makes renewable penetration faster and cheaper. Texas illustrated this beautifully when it established Competitive Renewable Energy Zones (described in a case study later on in this chapter), which was a planned strategy to build a suite of transmission lines connecting the windiest areas of Texas with load centers.

But even where access is available, it can be inhibited by overly complex interconnection standards, which in some jurisdictions seem designed to keep competitors out. The way around this is to have clear, straightforward interconnections and procedures, with reasonable time limits, for handling permits, and to apply nondiscriminatory standards. Some helpful design points for the interconnection process include reasonable but conservative screening to ensure an interconnection queue isn't crowded with projects that will not be built, a queue that prioritizes contracted projects over uncontracted projects for the same reason, a study process that allocates costs fairly among all projects in a given cluster, and a principle that upgrades cannot be added outside the official process and can be added only at specified times within the process.

Price Certainty Risk

The next major realm of uncertainty in developing a big renewable energy project is the price of the electricity generated. A long-term, highly certain price from a reliable purchaser makes it far easier to both invest capital at lower discount rates and raise competitive project financing. More generally, long-term power contracts also allow nonrecourse financing, which allows developers that don't have a large balance sheet to compete with those that do.

Utility-scale energy supplies have different sales conditions depending on the regulatory system in which they operate, and there is quite a range. Traditional vertically integrated monopolies—which own the power plants and transmission and distribution systems and sell directly to customers—can build their own power plants, or the utility may be the market maker, signing power purchase agreements with independent power producers. At the other end of the extreme lie competitive power markets where electricity is sold in 5-minute increments, with the help of a day-ahead auction. Both of these market structures are possible simultaneously because deregulated retail utilities sign contracts with different vendors to hedge and reduce risk. In California, projects can have long-term contracts but participate in the day-ahead or real-time market with a portion of the power plant output that is not contracted.

Clearly, a certain long-term price is more likely with a power contract than with a daily auction, but even in competitive power markets, bilateral contracting can lock up prices for a long while.

The ingredients for smart pricing are:

- For a vertically integrated monopoly building its own power plants, the regulator should allow a 10- to 20-year cost recovery schedule—but should certainly benchmark the price against those offered in competitive markets, so consumers do not overpay.
- For jurisdictions with the utility or the regulator as a market maker, offering solicitations to independent companies to build a new power plant, the regulator should offer 10- to 15-year contracts to bidders.
- For day-ahead and real-time competitive energy markets, the system should be structured to encourage a healthy volume of long-term bilateral contracts between energy marketers and energy suppliers.

In all cases, public incentives for clean energy should align with the time-scales needed for smart development (i.e., at least 10 years). Designing incentives over a long time horizon also helps avoid uncertainty about whether subsidies will be renewed, which can happen with subsidies that span only a few years at a time.

When to Apply These Policies

Regions around the world are undoubtedly in different stages of the move to clean electricity.[23] Context is very important for these policies. What does the current electricity mix look like? What power generation resources are available in the region, and how do their costs compare? The answer to these first questions can help sort which options for grid flexibility may be appropriate for the region.[24] Are regulated utilities, the government, or competitive providers the primary providers of electricity? The answer to this question can enable policymakers to dig deeper into performance-based regulation or well-functioning competitive power markets. Are all customers receiving adequate electricity service, or does the system need to expand to provide access to electricity for everyone? And is overall demand for electricity growing, staying steady, or falling? The answers to these last questions have a serious impact on which of these policies to prioritize and how, so this question is covered in more detail in the following sections.

Economies with Flat or Declining Electricity Demand

Flat or even declining electricity demand is increasingly common in developed countries. This is due in large part to major progress on energy efficiency policies (covered in other chapters) and improving technologies, such as light-emitting diodes (LEDs), which have resulted in a decoupling of economic growth from growth in energy demand. In regions with flat or declining electricity demand, all four mechanisms described earlier are important (support for grid flexibility, performance-based regulation, well-functioning competitive power markets, and orderly retirements), but orderly retirements may be especially important.

Electricity systems that historically relied on coal, oil, or natural gas will need to see retirements in order to make room for zero-carbon power. This fact creates losers among power generators, which may have unpaid capital balances left on their investments. Because of the dominant role of fossil-fueled power in the past, early retirement of these plants is likely to require new approaches to running the grid, and the changes will prompt calls for support for fossil-based resources. Policymakers should objectively assess whether adequate clean replacement generation, coupled with cost-effective efficiency, storage, and demand response, is available to keep the lights on as old power plants retire.[25] If so, power sector regulators should continue guiding smart retirements and replacements while finding ways to support workers as they transition from old industries.

Economies with Growing Electricity Demand

Regions may see electricity demand growth for several reasons: Perhaps the overall population is growing, energy-intensive industries are growing, more people are emerging from poverty and able to afford electricity, or perhaps electric vehicles are experiencing substantial growth. Whatever the source, regions with consistent electricity demand growth should look seriously at strong renewable portfolio standards or feed-in tariffs, as well as all four mechanisms described in this chapter.

Growing demand can signal macroeconomic trends, but it can also signal opportunities to improve energy efficiency and warrant higher standards for buildings and industry. Efficiency investments are usually the most cost-effective zero-carbon resource to meet growing demand,[26] and economies with growing electricity demand should closely examine performance standards such as energy efficiency resource portfolios and efficiency incentives for their utilities.

Of course, orderly retirements will be less of a focus in a high-growth environment, but it is nevertheless a sign of a healthy power sector when clean generation sources begin to replace older, polluting power plants at the same or lower cost for customers. Orderly retirements may still be proper when existing resources are dirtier and more expensive to run than new clean energy replacements, as long as sufficient financing is available to both meet growing demand and replace old generators.

Economies without Universal Access

Deploying and maintaining electricity infrastructure is a major challenge for developing economies that have not yet achieved universal electricity access for their citizens. Expanding access by connecting customers to the existing grid often requires building massive distribution and transmission infrastructure projects, necessitating large amounts of capital that may not be readily available. High upfront costs coupled with potentially low returns from customers who may have trouble paying their bills consistently can undermine the ability to attract capital for infrastructure buildout, often necessitating some level of state support. Even where grid connections have been established, reliability of service can be low, undermining the perceived value of grid expansion in a vicious cycle.

Many off-grid households rely on smaller, distributed fuel-based methods of generating power or heat, such as diesel generators or burning coal, oil, or animal dung to cook, light, or heat buildings, but those are heavy polluters, harming community health. Fortunately, technological developments have enabled new options for clean electricity, many of which can actually save customers money relative to fuel-based alternatives.

With technology costs declining, novel options for energy access are emerging, ranging from super-efficient solar home systems to community microgrids. Off-grid or community-based shared solar and battery arrangements provide access to electricity at a lower cost and can be deployed more nimbly than grid infrastructure extension in many cases. The Sierra Club estimates that rural access to energy via solar-powered minigrids will cost about $250 per customer, whereas grid extension costs about $1,000–$2,500 per customer, depending on the distance from the existing grid.[27]

In the context of providing electricity access, policies should focus on mobilizing capital and building institutions to enable access rather than refining large institutions through the changes to utility regulation and wholesale market rules described earlier. Some of the Sierra Club's principles to expand access to clean electricity in off-grid systems are particularly salient:

1. Support deployment and development of highly efficient appliances and agricultural equipment.
2. Focus on providing small investments that provide basic access, then build financing support as markets mature and incomes increase to pay for more services.
3. Reduce costs of access by eliminating private investment risk through loan guarantees or rural feed-in tariff subsidies that guarantee cost recovery.
4. Define utility regulations in the off-grid and minigrid space.[28]

Detailed Design Recommendations

Policy Design Principles
Several policy design principles apply to policies that facilitate a low-cost transition to renewable energy in the electricity sector.

Create a Long-Term Goal and Provide Business Certainty

Performance-based regulation works best when performance targets are stable and extend well into the future. Investor-owned utilities are viewed as low-risk enterprises that are able to pass on the benefits of low-cost capital to their customers (low risk is ultimately reflected in lower electricity prices). Under cost-of-service regulation, most of the risk to utility investors is due to regulatory uncertainty—regulators who have approved capital investments at one time may change, increasing the chance that the utility will not be allowed to collect the full cost of its investment in the future. Subjecting a portion of utility returns to performance-based regulation introduces additional risk when there is a lack of certainty as to whether the utility can meet regulation targets. However, performance-based incentives should have an upside as well, creating opportunities to increase returns and offset higher risks.[29]

Any change to the utility compensation model creates some uncertainty, but this can be mitigated through smart policy design. The first principle should be to set performance goals over a long enough time horizon, at least 5 but up to 8 years. Longer timelines provide the business certainty needed for innovation, but targets set too far into the future can be overwhelmed by uncertainty about exogenous trends or events. Whatever timeline regulators choose at the beginning of a program, they should not materially alter the goals until the chosen performance period is over, in order to support a stable business environment and avoid the perception of greater regulatory risk that unnecessarily raises utilities' cost of capital. In the context of revenue cap regulation, utilities and their investors need confidence that regulators will stick to a revenue cap even if utility profits increase, maintaining the efficiency incentive.

Long-term policy certainty is also crucial for de-risking renewable energy development. One example of the need for policy certainty is the U.S. production tax credit (PTC) for renewable energy. Although the PTC was fixed for 10 years, the U.S. Congress allowed it to lapse five times and delayed extension to the final days of the year in other years, causing turmoil and uncertainty in the wind industry, even driving some producers out of the market. Without certainty that the PTC would be extended, the wind industry built incredibly fast in the on years and ceased construction in the off years.[30] This boom–bust cycle of investment is hard for companies to maintain and leads to inefficient outcomes (Figure 5-4).

One could be forgiven for thinking that these policies—on-again, off-again credits with limited liquidity—were designed to vex the very industry they were supposed to help.

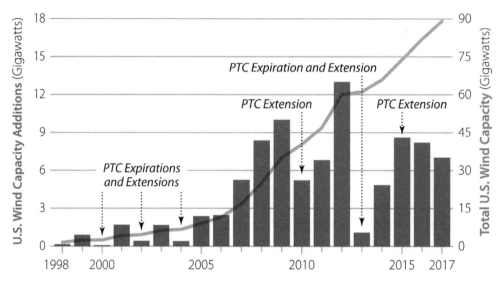

Figure 5-4. Policy uncertainty creates wind energy boom-and-bust cycles. (From Ryan Wiser and Mark Bolinger, "2016 Wind Technologies Market Report," U.S. Department of Energy, 2016, https://energy.gov/sites/prod/files/2017/10/f37/2016_Wind_Technologies_Market_Report_101317.pdf. Data taken from EIA Wind Technologies Report 2016, https://energy.gov/eere/wind/downloads/2016-wind-technologies-market-report.)

The United States largely fixed this problem with the 2016 extension and gradual phase-down of renewable energy tax credits. The wind PTC phases down to zero gradually from 2017 to 2020, and the investment tax credit for solar technologies phases down incrementally from its 2017 level of 30 percent to 10 percent in 2022. A gradual phase-down, announced in advance, for technologies close to market competitiveness can also achieve a de-risking effect.

Price In the Full Value of Negative Externalities
Wholesale electricity markets send the wrong price signals when externalities are not reflected in the marginal price of generation. This is particularly true of carbon and other air pollutants that affect human health. Because centralized wholesale markets rely on the marginal price of energy to dispatch different generators in a least-cost fashion, what may appear to be cheap coal or natural gas is costing more than it appears when only direct fuel costs are considered. In the absence of a carbon price in the region or at the fuel source, wholesale

markets can consider adding a proxy social cost of carbon calculation to the bids of generators based on their carbon intensity.[31] This will create more revenue for zero-marginal-cost resources such as wind and solar while promoting zero-emission dispatchable resources such as nuclear and hydro over their main competitors: natural gas, coal, and oil.

Eliminate Unnecessary Soft Costs

Lowering development risk for renewables is a key measure to reduce soft costs—the nonhardware component of a project's costs—when building wind and solar projects. Good policy can drastically cut this uncertainty by pre-zoning land and setting clear requirements and timeframes for permits. For example, a public land agency such as the Bureau of Land Management can zone land as green, yellow, and red. Green zones would be suitable for renewable energy development as long as clear, prespecified conditions are met. Red lands would never be developed; don't even ask. Yellow lands would open up a complex, careful process to determine whether they are suitable for development. Today, almost everything is, de facto, yellow and vastly inefficient.

This is emphatically not an argument for relaxed environmental standards. Policymakers and communities protect viewscapes, landscapes, habitat, wetlands, streams, watersheds, and species for good reason. Instead, this prezoning offers an efficient way to land at the right result. The red-zoned lands would include wilderness study areas, for example. Conversely, a green zone might be an existing oil and gas field, or an interstate highway corridor, or open land that is not critical habitat.

Permitting can also be streamlined while protecting environmental standards. The first step is to enumerate the required steps a project developer must take—and to do so across all agencies that have a say in the matter, be they federal, state, or local. The work of identifying the current permit critical path should be done hand in hand with experienced developers, because they are more likely than government officials to understand the full picture.

The work should then be collated by purpose and by jurisdiction and a clearer, simpler set of requirements produced. The result should:

- Use common forms wherever possible. If colleges can use a standard application form, there is no reason that states and local agencies can't handle overlapping requirements the same way.
- Be crystal clear on what is required for a successful permit. The goals of

the standards need not be relaxed, but they should not be ambiguous, and it should not be ambiguous what it takes to meet the standard. There should be clear timelines, both in terms of submission by developers and response by the permitting authority so that timeline clarity is evident. This will allow developers to appropriately raise and invest capital according to when key risks (such as binary or discretionary permit approvals) are mitigated.

- Reduce or eliminate any processes that are not necessary. Ideally, build a master file system so that applicants do not need to fill out redundant information in dozens of places. Where possible, commit to a paperless workflow process.
- Set a firm, quick time commitment for approval of any permit that meets all the requirements.
- Commit to a reasonable response time for any issue that arises.
- Good policy should be developed and shared between jurisdictions (e.g., in California, between counties) so that reasonable standards are developed and are replicable across multiple locations, meaning developers will not have to reinvent the wheel in every new jurisdiction they enter. This allows them to cost-effectively deploy resources and import best practices into more areas.

A complementary government program could be to appoint an ombudsman who, with policymakers' blessing, can help developers clear through regulatory and permitting problems.

With this process, crafted with care and published, developers would know exactly what the requirements are for developing and permitting a new solar farm. They would have much more clarity for timing, costs, selecting contractors, and so forth, which would help investors understand that this is a lower-risk project based on clear timelines and steps to approval.

Build In Continuous Improvement
Performance-based regulation is effective when the performance standards set by regulators build in continuous improvement. The state of the art for electricity service is constantly evolving as technologies enable a more connected, responsive grid. It is now possible for system operators to have visibility into the system and predict, adjust to, and avoid reliability issues in ways that were

not possible just 10 years ago. Performance targets should build in continuous improvement to reflect this reality, such as by setting targets as annual improvements rather than hard numbers tied to a specific future date.

This principle goes hand in hand with creating business certainty, because the targets for performance under performance-based regulation are also linked to utility compensation. Creating performance targets with continuous improvement also creates long-term certainty that ensures businesses will have the right signals and time horizons to make strategic investments that meet public policy goals while preserving utility profitability.

Focus Standards on Outcomes, Not Technologies

Centralized wholesale markets and vertically integrated utilities should be technology neutral in the way they achieve low-cost, low-carbon electricity service. In the wholesale market context, this applies to which resources can participate in market transactions. Many wholesale markets create artificial barriers to clean technologies in the way they adopt standards or define products. For example, participation in a market for ancillary services such as frequency response may require a synchronous generator, making it impossible for inverter-based generation technologies such as wind, solar, and battery storage to participate in and receive revenue from those markets, even though they are technically capable.[32] As markets contemplate ways to value much-needed flexibility to balance variable renewables, they should remain technology neutral, allowing different combinations of resources to provide the lowest-cost set of grid services needed.

In the vertically integrated context, utility bias toward conventional, capital-intensive investments can result in an unjustified preference for large fuel-fired generators and conventional grid infrastructure over demand-side options, storage, and renewables. Performance-based regulation can help encourage utilities to become technology neutral in their selection of supply- versus demand-side resources to provide cost-effective service to customers. In particular, a revenue cap and incentives for efficiency can be particularly powerful drivers to incentivize technology neutrality.

Prevent Gaming via Simplicity and Avoiding Loopholes

Gaming in the utility performance context is explained in a handbook on performance incentive mechanisms that Synapse Energy Economics prepared for regulators:

Every performance incentive mechanism carries the risk that utilities will game the system or manipulate results. "Gaming" refers to a utility taking some form of shortcut in achieving a target so that the target is reached, but not in a way that was intended. For example, if a performance incentive were set that rewarded a utility for increasing a power plant's capacity factor above a certain threshold, the utility might understandably respond by increasing its off-system sales from that power plant, even at an economic loss. Thus the utility would be able to meet or exceed the target capacity factor, but ratepayers would be worse off.[33]

A key in designing performance targets and incentives is to tie them as closely to outcomes as possible. For example, if the goal is to improve system efficiency, regulators might consider measuring and setting a target for peak demand reduction rather than requiring the utility to spend some fixed percentage of its budget on demand response or buy a fixed amount of storage. The utility could meet these targets without verifying that these technologies are being used to actually reduce peak demand. Instead, the metric of peak demand reduction can be measured and verified by observing the peak demand in a start year and subsequent years.[34]

Performance standards or targets under performance-based regulation can be gamed by utilities if they are not designed transparently. The first key is to create transparency on how utility performance is measured. This means the methods and the underlying data should be shared with all stakeholders, who should be able to replicate the utility's performance calculation using the same publicly available data. The second principle is simplicity. The data collection and analysis techniques should be straightforward if possible, enabling regulators and other stakeholders to more easily determine the data's accuracy. This in turn makes manipulation of data more difficult and reduces the costs of oversight.[35]

Additional Design Considerations

Regional Coordination

One of the most cost-effective ways to integrate high shares of renewable energy is to increase geographic diversity of the resources being balanced. Over larger distances, wind patterns are negatively correlated, meaning one place will be windy when the other is not. This reduces the need for backup generation and storage to even out wind's variable production profile.[36] Likewise, as the sun moves east to west over a continent, importing solar from the west can help manage local sunset ramps.[37]

Bigger grid balancing areas can be achieved several ways, including merging smaller balancing areas or simply allowing trading of electricity between existing balancing areas. For example, special markets are developing in the United States to trade grid balancing services between regions that are operated independently from one another. Without needing to build new physical transmission capacity—simply by allowing trades between regions that did not allow them before—these burgeoning markets have saved customers in the western United States at least $140 million per year.[38]

In addition, greater transmission connectivity can help increase trading capacity and increase the diversity of demand and supply options to manage variable wind and solar generation. Although transmission is expensive and difficult to site, planning renewable generation and transmission projects together can yield tremendous efficiencies.[39] One report that studied the U.S. grid found that the grid could run reliably and at the same cost as in 2013 on more than 50 percent wind and solar generation, reducing greenhouse gas emissions by 80 percent from 2005 levels, but this required a coordinated buildout of high-voltage direct current transmission to allow trading to access high-value renewable resources and mitigate variability.[40]

Publicly Owned Utilities

Unlike investor-owned utilities, which are ultimately motivated by profits and shareholder value, publicly owned utilities are nonprofit entities that are owned by customers themselves, through either the state or local government or a cooperative arrangement. Publicly owned utilities are directly connected to public policy and governed democratically. The transition to high shares of renewables is just as challenging for public utility management and boards as it is for their investor-owned utility counterparts, but publicly owned utilities may not be affected by the specific recommendations for wholesale energy market reform or performance-based regulation covered in this chapter.

However, some principles of performance-based regulation can be applied to encourage innovation and measure progress at publicly owned utilities. Consistently setting, measuring, and updating quantitative performance metrics, particularly around demand-side management and carbon intensity, should be a central feature of any publicly owned utility management program.[41] For small utilities this can be a large lift, but even the simplest goals are useful places to start orienting utility operation around increasing customer value.

Publicly owned utilities are also similar to investor-owned utilities in the way in which energy efficiency can hurt their bottom line. Energy efficiency reduces the total volume of electricity sold by avoiding some demand. To recover the costs of past investments, utilities set rates based on the expected amount of electricity they will sell. If a utility then implements energy efficiency measures and lowers the amount of electricity it sells, it will not recover the full costs of its past investments. One way around this problem is to pursue revenue decoupling. Decoupling allows utilities to retroactively recover any lost revenue or return surplus revenue to customers, based on the amount of electricity sold. It therefore eliminates the incentive for utilities to sell more electricity.[42] Publicly owned utilities worried about their revenue due to falling sales can consider decoupling as a way to reduce financial uncertainty and drive energy efficiency.

Refinancing Options for Coal Retirements

Transitioning uneconomical coal- and oil-fired power plants off utility balance sheets can be much cheaper if stakeholders take advantage of ratepayer-backed, state, or municipal bonds with very low interest rates. As explained earlier, many uneconomical plants have undepreciated balances, and early retirement would require the owners to take that balance as a loss. Moving the balance out of the utility rate of return structure and into a low-cost bond can significantly reduce the costs of paying off these old plants for customers.

The typical investor-owned utility's authorized return—its weighted average cost of capital—reflects a mix of corporate bonds and shareholder equity. The weighted average cost of capital, in turn, depends a lot on the financial viability of the utility, its ability to collect all its costs from customers, and the investment climate and borrowing costs in the country in which it exists. In general, the cost of equity exceeds the cost of debt, and government-backed debt costs even less than corporate debt. As a result, shifting the balance of plant costs from a utility balance sheet, where they receive the weighted average cost of capital, to one backed by a government or by customers in aggregate, can reduce financing costs, sometimes by more than 50 percent.[43] This in turn makes it attractive for utilities to refinance, and ultimately retire, plants.

Case Studies

United Kingdom

The UK's utility regulator, the Office of Gas and Electricity Markets (Ofgem), uses price controls or revenue cap regulation plus incentives to drive efficiency in regulated utilities by setting revenues for a long time and letting the companies keep a profit if they manage to deliver the same outputs at a lower cost. This program, called Revenue = Incentives + Innovation + Outputs (RIIO), is the most full-scale move toward performance-based ratemaking for utilities observed anywhere in the world to date.

At the heart of RIIO are detailed "business plans" that each utility in the UK must develop and submit to the regulator. The business plans specify the expenses each utility expects over the next 8 years, forming the basis of the revenue cap. Utilities that reduce expenses below the revenue cap are able to keep about half the savings as profits, sharing the other half with customers. Utility investors also share in excess expenses as losses, producing strong incentives for cost efficiency.

To ensure a high level of service is maintained, the business plans must also be responsive to performance incentives in six primary output categories: customer satisfaction, reliability and availability, safe network services, connection terms, environmental impact, and social obligations—each with measurable targets. Achievement of many of these targets results in higher overall profit for the utilities. Failure to meet the targets can sometimes mean a penalty for the utilities. For example, Ofgem proposed to add or subtract a maximum of half a percent of revenues based on a customer satisfaction scoring system.

The dual impact of a long-term revenue cap and long-term performance incentives creates business certainty and builds in continuous improvement. According to Figure 5-5, all utilities were able to earn returns above their cost of capital in the first performance period, creating value for shareholders. Moreover, the utilities performed well on the metrics that were linked to performance, indicating that the incentives promoted real-world improvement.

There may have also been some performance targets that were too loose, yielding windfall profits for the utilities in the first performance period. Although some UK utilities are paying penalties for underperformance on outcomes, most are successfully earning incentives for performing well on outcomes. This mix of penalty payments and incentive earnings is balanced, but

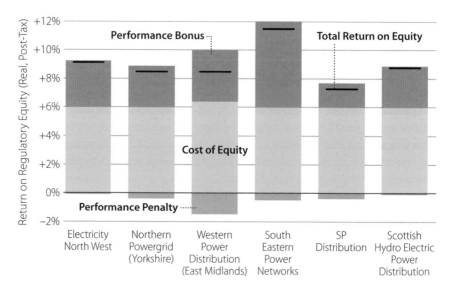

Figure 5-5. Performance-based regulation can yield total returns on equity above the cost of equity. (Data reproduced from publicly available UK government data under the terms of the Open Government License from "RIIO-ED1 Annual Report 2015–2016," Ofgem, 2017, https://www.ofgem.gov.uk/system/files/docs/2017/02/riio-ed1_annual_report_2015-16.pdf.)

it is worth noting that more utilities are performing (and earning) well than poorly on their outcomes. This may indicate that more ambitious performance targets could have been warranted to better share benefits between utilities and customers or that utilities were able to take advantage of information asymmetry (knowing more details about performance potential and associated cost than their regulators).

Texas

Texas houses one of seven centralized wholesale markets in the United States, the Electric Reliability Council of Texas (ERCOT). ERCOT is unique among U.S. markets in that it operates over the footprint of a single grid, whereas the others in the United States operate in regions that are parts of a larger, interconnected grid. Texas's wholesale market has several other unique features: It operates without a market for capacity, and it has a very high cap on the price of energy, at $9,000 per megawatt-hour.

Texas has also been one of the most successful integrators of wind power, with 21,000 megawatts of capacity where peak load is 71,000 megawatts.[44] In

2016, 15 percent of ERCOT's generation was produced by wind power; in March 2017, a record 25.4 percent of total generation was from wind power.[45] ERCOT's policymakers have made two important recent policy decisions that have brought great benefits to their electricity consumers while driving a cleaner grid.

The first of these, completed in 2013, was the creation of a set of competitive renewable energy zones that funded nearly $7 billion[46] in transmission projects to tap renewable energy resources in windy, remote west Texas (Figure 5-6).

The lines installed in these zones immediately relieved major transmission bottlenecks and gave a clear market signal to renewable energy project developers, enabling more than 20,000 megawatts of new wind, triple the amount of the next highest state (Figure 5-7).[47]

The zones are technology neutral and do not require access to wind;

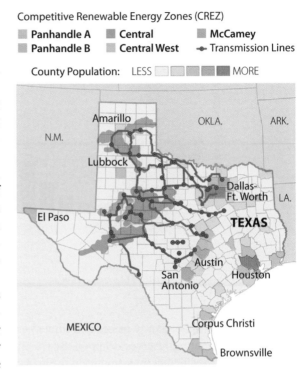

Competitive Renewable Energy Zones (CREZ)

- Panhandle A
- Panhandle B
- Central
- Central West
- McCamey
- Transmission Lines

County Population: LESS ▢▢▢▢■ MORE

Figure 5-6. Competitive Renewable Energy Zones in Texas. (From "Fewer Wind Curtailments and Negative Power Prices Seen in Texas after Major Grid Expansion," U.S. Energy Information Administration, 2014, https://www.eia.gov/todayinenergy/detail.php?id=16831.)

rather, they have enabled wind resources to participate in the market. It is anticipated that they will also stimulate at least 3,000 megawatts of future solar projects[48] with complementary generation profiles to wind (which blows mostly at night in west Texas). They have also reduced soft costs for renewable developers, creating access to world-class wind resources without risk and delays created by uncertain transmission access. As a result, new low-cost wind is projected to offset the costs of the program and associated transmission lines and then some, saving customers an estimated $16 billion through 2050.[49]

The other important decision recently made by Texas policymakers was to forgo the use of a capacity market as a means to ensure that enough generation would always be there to serve load. Instead, Texas raised its caps on

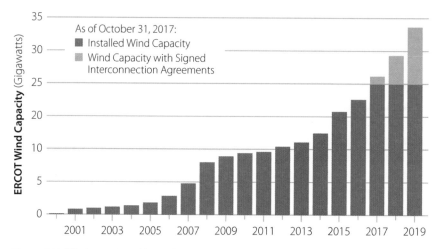

Figure 5-7. Wind capacity additions by year in Texas. (Data used with permission from the "ERCOT Generator Interconnection Status Report" [ERCOT, 2017], http://www.ercot.com /content/wcm/lists/114799/GIS_REPORT__October_2017.xlsx.)

real-time energy prices and implemented a price adder that would kick in if operating reserves dropped. By trusting a well-designed wholesale market to elicit efficient investments in the grid, Texas customers avoided costly capacity payments on the order of billions of dollars.[50] Yet ERCOT was able to retire thousands of megawatts of inefficient old gas and coal power plants while taking advantage of record low prices for new gas and wind to enable a cleaner, cheaper, more flexible grid with record reliability.[51] The fact that ERCOT has more than enough capacity provides evidence that a capacity market may not be necessary to provide long-term business certainty sufficient to stimulate investment in adequate power generation.

By running a very efficient, technology-neutral wholesale market, Texas has been able to reap the benefits of its renewable portfolio standard and investment in the transmission lines. Smart policies were able to jump-start a wind (and soon solar) market that will be delivering benefits to the citizens of Texas for years to come.

Germany
Germany has also seen major growth in renewable energy in part because of its policies to reduce the investment risk of renewable energy. Policies to reduce

investment risk have reduced soft costs in Germany, provided long-term certainty, and built in continuous improvement. Of course, some of the other supporting policies described in this chapter are missing, which is part of the reason it is taking Germany a little longer than other regions to realize the emission reduction benefits of its renewables.

Financing is a large contributor to soft costs, and Germany has some of the lowest financing costs for wind and solar in the European Union. Typical financing costs are between 3.5 and 4.5 percent for onshore wind projects.[52] In large part, Germany's success in driving down financing costs results from the availability of low-cost capital from state-owned development banks. German development banks are able to provide loans at 2 or 3 percent interest, taking advantage of the creditworthiness of the German government. Domestic developers finance between 80 and 100 percent of onshore wind projects with low-cost debt, using a low share of more expensive project equity.[53] Climate Policy Initiative estimates that between 60 and 70 percent of the total funding for renewable energy investment in Germany in 2013 and 2014 was originally provided by development banks.[54]

Long-term certainty and continuous improvement have also greatly derisked renewable investment in Germany. Feed-in tariffs are structured with long-term contracts (8–15 years) at locked-in rates, allowing investors to recover the majority of their investments with almost no risk. From 2002 to 2014, the feed-in tariff had minimal price risk, as renewable generators received fixed prices; as the market matured, this was changed in 2014 to a feed-in premium structure under which generators receive a reduced feed-in tariff on top of market-based electricity prices.[55] Germany's 2004 renewable energy goal had targets of 12.5 percent by 2012 and 20 percent by 2020 (Figure 5-8).[56] That goal was increased in 2012 to 35 percent by

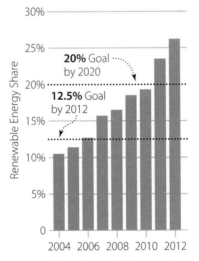

Figure 5-8. Share of renewable energy in Germany, 2004–2012. (Analysis done using data with permission from the Fraunhofer Institute, "Energy Charts," accessed December 12, 2017, https://www.energy-charts.de/power_inst.htm.)

2020, 50 percent by 2030, 65 percent by 2040, and 80 percent by 2050, providing policy for continuous improvement over 38 years.

Conclusion

Achieving an electricity system with high shares of variable renewable energy will be economically feasible only if complementary policies evolve with renewable energy standards. In particular, to create a more flexible electricity system that supports much more wind and solar, regulators will need to make deliberate choices about how utilities make money, how wholesale markets operate, and how we can transition away from existing fossil fuel generation. Without these tools, the cost of the renewable energy transition may be higher than it needs to be, creating roadblocks to economic growth, shutting out energy-poor communities, and reducing the competitiveness of energy-intensive industries.

THE TRANSPORTATION SECTOR

The transportation sector is responsible for more than 15 percent of annual global greenhouse gas emissions, with the most recent data showing carbon dioxide emissions of about 7.5 billion tons in 2014.[1] Emissions are expected to grow to more than 9 billion tons by 2050.[2] Without additional policies, the transportation sector will be responsible for 14 percent of cumulative emissions through 2050.[3]

The growth in emissions is largely due to increasing car ownership and freight transport. For example, passenger travel demand is expected to more than double between 2010 and 2050, and freight transport is expected to increase by nearly 60 percent over the same period.[4] Without additional policy, the vast majority of this demand will be met with petroleum fuels, causing emissions to grow.

Reducing emissions from the transportation sector requires improving the efficiency of vehicles produced and the average efficiency of vehicles sold, increasing the share of electric vehicles sold, and providing alternatives to owning and driving a vehicle through smart urban planning. Chapter 6 looks at how vehicle performance standards can improve the efficiency of vehicles in the market. Chapter 7 looks at vehicle and fuel fees and feebates and how they can encourage consumers to purchase more efficient vehicles and drive them less often. Chapter 8 evaluates policies that can promote electric vehicle adoption. Finally, Chapter 9 explores urban mobility and how policymakers can design and encourage investment in low-carbon cities with plenty of transportation options.

Decarbonizing the transportation sector is an important element of any

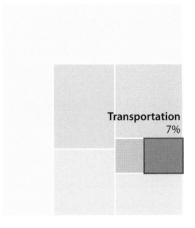

Figure S-2. Potential emission reductions from transportation sector. (Analysis done using data with permission from the International Institute for Applied Systems Analysis [IIASA]. Data source: Tavoni et al., 2013. Data downloaded from the LIMITS Scenario database hosted at IIASA, https://tntcat.iiasa.ac.at/LIMITSPUBLICDB /dsd?Action=htmlpage&page=about.)

climate strategy, with significant co-benefits such as reduced particulate pollution and lost time due to traffic. Together, the transportation sector policies discussed here can contribute at least 7 percent of the reductions needed to meet the two-degree target (Figure S-2).

CHAPTER SIX

Vehicle Performance Standards

In 2014, the transportation sector emitted 7.5 billion tons of carbon dioxide equivalent (CO_2e), accounting for more than 15 percent of global greenhouse gas emissions.[1] Emissions from road transportation are projected to grow more than 30 percent by 2030,[2] spurred largely by a dramatic increase in cars and trucks in China, India, and other developing economies. Although on-road vehicles account for the lion's share of transportation sector emissions (72 percent), aviation (11 percent) and shipping (11 percent) are also important contributors,[3] and their share of transportation emissions is likely to rise as on-road vehicles become more efficient.

The policies with the best track record of significantly reducing fuel consumption and lowering emissions from vehicles are vehicle performance standards. Vehicle performance standards are a powerful tool for improving the fuel efficiency of and reducing greenhouse gas emissions from newly sold vehicles. Standards should be known by manufacturers years in advance, giving them time to adapt their product offerings, and should continuously improve over time. They should be technology neutral and written simply, reducing opportunities for gaming. Japan's Top Runner program, China's Vehicle Emissions Control Program, and the U.S. Corporate Average Fuel Economy standards include well-designed elements, although none is an unmitigated success. Designed well, stronger vehicle performance standards can achieve about 3 percent of cumulative global emission reductions needed to meet the two-degree target (Figure 6-1).

Policy Description and Goal

Vehicle performance standards, often called fuel economy standards, establish maximum allowable levels of fuel consumption or greenhouse gas emissions per unit distance traveled. Their objective is to ensure that all vehicles, either

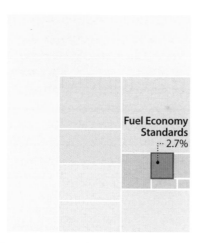

Fuel Economy Standards
∷∙ 2.7%

Figure 6-1. Potential emission reductions from vehicle performance standards. (Analysis done using data with permission from the International Institute for Applied Systems Analysis [IIASA]. Data source: Tavoni et al., 2013. Data downloaded from the LIMITS Scenario database hosted at IIASA, https://tntcat.iiasa .ac.at/LIMITSPUBLICDB/dsd?Action=htmlpage &page=about.)

individually or as a weighted average of sales, achieve certain minimum performance targets, thereby raising the vehicle fleet's energy efficiency and reducing emissions. Past vehicle performance standards have reaped outstanding benefits: lower oil consumption, lower greenhouse gas emissions, reduced dependence on oil imports, and increased investment in innovative technologies, all at a net savings to society. Vehicle performance standards are particularly valuable because market barriers may limit the efficacy of other transportation sector policies, such as vehicle and fuel fees and feebates (discussed in Chapter 7), in delivering deep energy and emission reductions.

Vehicle performance standards have been implemented in a variety of ways in different countries. When designing the policy, some of the major choices to be made are:

- How to differentiate the standard for different vehicle types (trucks, cars, etc.).
- Whether to cover all types of vehicles under the standard or whether to exclude certain vehicle types, such as two-wheeled or three-wheeled vehicles. Generally, all vehicles should be covered.
- Whether the standard should control the amount of fuel consumed or the amount of greenhouse gases emitted per unit distance traveled. For many jurisdictions, standards should ideally be emission based, which better accounts for leaking refrigerant from air conditioning systems and allows the use of other control technologies, which are relevant for non-CO_2 pollutants. However, some jurisdictions may prioritize oil conservation and create a standard intended primarily to reduce fuel consumption.

- Whether the standard will make distinctions based on vehicle weight, vehicle footprint, or other design factors. If any distinction is required, it should be based on footprint (its two-dimensional projected area on the ground).
- Whether the standard has built-in, automatic improvement, or whether it must be manually reviewed by a legislative or regulatory body for each update. Ideally, standards should have a built-in improvement mechanism.
- Whether the performance standard in a given year is defined as a continuous function for each vehicle type (with respect to an independent variable reflecting the vehicle's design, such as footprint) or whether performance targets take the form of discrete stairsteps.
- Whether to apply a standard to every vehicle sold or to apply the standard as a sales-weighted average of vehicles of a given type sold by a particular manufacturer. A sales-weighted average allows more flexible compliance options. A hybrid option is possible as well. For example, China sets both fleet-wide average standards and a minimum requirement for each model in a specific class.[4]

When to Apply This Policy

Using vehicle performance standards is almost always smart policy: They save society money while driving down emissions. However, there are particular policy objectives for which these standards are well suited and others that are not well suited, leaving a role for complementary transportation sector policies such as fuel fees and vehicle feebates, which are discussed in the next chapter.

When to Apply Vehicle Performance Standards

When standards specify minimum performance requirements that every vehicle must meet, they affect the lowest (worst-performing) segment of the marketplace. They force manufacturers to either improve their vehicles or stop selling the most polluting vehicle models and instead sell their better-performing vehicles. This approach has advantages. For example, efficient vehicles already exist in the marketplace for virtually every vehicle class. Most manufacturers already produce models with a range of performance characteristics, which demonstrates the feasibility of complying with the standards. Therefore, standards are a particularly good fit for regions where some vehicle manufacturers sell significant numbers of inefficient vehicles, whether due to older technology or by favoring larger vehicles, such as SUVs.

When standards are specified as minimum performance requirements on a sales-weighted average basis (for each vehicle type, for each manufacturer), manufacturers may improve the worst-performing vehicles they sell, or they may improve the best-performing vehicles and sell more of them (offsetting the emissions from the worst-performing vehicles), or both. This helps guide not only manufacturers' design choices when developing or refining new vehicle models but also their marketing and pricing choices, to ensure the sales-weighted average hits the target. If vehicle manufacturers that overcomply on a sales-weighted average basis are allowed to sell this credit to other manufacturers who may then legally undercomply with the standard, the policy gives efficiency leaders an incentive to innovate further so as to have more credits to sell to other manufacturers. This design is good for countries or regions that can establish, monitor, and prevent cheating on a credit trading marketplace.

Another factor to consider when deciding to apply vehicle performance standards relates to overcoming market irrationalities and barriers. Although evidence is mixed, many studies have found that consumers significantly undervalue fuel savings when making their purchasing decisions.[5] Unlike economic incentives (taxes or subsidies), a performance standard does not rely on consumers making rational adjustments to their behavior in response to an economic signal. A performance standard will work irrespective of whether consumers care about fuel efficiency or even in the case where consumers buy vehicles without knowledge of their fuel efficiency. This makes standards good fits for countries or regions with buyers who have limited information or are insensitive to fuel efficiency.

Vehicle performance standards are additionally important in the face of fuel price uncertainty. Consumers might be likely to undervalue fuel savings, but they also often don't have good information about fuel prices, which can fluctuate drastically even within the timeframe it takes to make a vehicle purchase decision. Manufacturers and consumers alike benefit from standards that prepare and protect them from volatile oil prices.

When to Apply Complementary Policies

Several worthy policy goals are not tackled by vehicle performance standards. First, standards affect only new vehicles; they do not encourage the retirement or replacement of old, inefficient vehicles. In practice, this means vehicle performance standards have a long lead time after the policy is enacted before their full emission benefits are realized. A policy to encourage the early retirement of old vehicles—such as government buyback and scrapping of inefficient vehicles ("cash for clunkers")—may be a helpful complement that accelerates the impact of vehicle performance standards.

Second, among vehicles that comply with the standard, the policy provides no incentive for a buyer to select one of the best-performing options rather than one that narrowly complies. A performance standard is weaker than some other policies at pushing consumers to adopt cutting-edge technology and at encouraging manufacturers to produce innovative, extremely efficient new models. That said, fleet average standards that allow manufacturers to sell a mix of high-performing and low-performing vehicles are likely to result in more of an innovation push than standards that simply set a minimum requirement per vehicle, especially if less efficient vehicles are cheaper or more likely to sell. Additionally, in some systems overcomplying manufacturers may sell credits to other manufacturers, meaning some manufacturers can just buy their way out of compliance. A pricing policy that provides financial rewards that increase with vehicle efficiency, such as a feebate (discussed in Chapter 7), helps mitigate this weakness.

Third, a vehicle performance standard does not reduce the amount a vehicle is used. In fact, by lowering the cost of travel, a vehicle performance standard can actually encourage an increase in vehicle use. Studies vary in estimating this effect, but it is reasonable to conclude that a 10 percent increase in vehicle efficiency results in roughly a 2 to 4 percent increase in vehicle use. Although the rebound effect is not trivial, it is more than offset by savings from the fuel economy standards. For example, assuming an average distance traveled of 12,000 miles per year, increasing a vehicle's efficiency from 25 miles per gallon to 30 miles per gallon would result in a reduction of 80 gallons per year without the rebound effect and 73 gallons per year assuming a rebound effect of 10 percent. Although the rebound effect reduces the amount of fuel saved by about 7 gallons per year, the net savings of 73 gallons vastly outweighs this effect.

Lastly, policies governing planes and ships, often used for international transport, tend be set internationally. The International Civil Aviation Organization has recently set the first-ever vehicle performance standards for aircraft,[6] and the International Maritime Organization is doing something similar for oceangoing vessels.[7] In these cases, rather than setting their own vehicle performance standards (and generating a patchwork of regulations), policymakers can adopt and enforce the standards set by these organizations. As members of these groups, countries may push these bodies to enact strong, well-designed standards. Aviation and marine emissions are already a significant source of emissions, and addressing emissions from these sources will become increasingly important as the rest of the transportation system becomes electrified and decarbonized.

Detailed Design Recommendations

Policy Design Principles

The following policy design principles apply to fuel economy standards.

Create Long-Term Certainty to Provide Businesses with a Fair Planning Horizon

Vehicle and auto manufacturers may need several years to make the investment in research and development (R&D) necessary to meet fuel economy standards. If standards are set only a few years at a time, businesses may be unsure whether standards will continue to become more stringent in the future. In turn, they will not know whether modest investment in near-term improvements is sufficient or whether a larger investment in a new technology or major design change might be worthwhile.

Because of their greater size and complexity, the design cycle for aircraft and oceangoing ships is even longer than for automobiles, so a long-term planning horizon is even more crucial for these industries.

A known schedule of vehicle performance standards helps companies justify investments in R&D and fuel efficiency to their shareholders, who otherwise might be skeptical of the value of these expenditures. Companies that make a serious commitment to R&D may see tighter standards as their competitive advantage.

Build In Continuous Improvement

Vehicle performance standards are valuable because they continue to drive the vehicle fleet toward greater efficiency and eliminate the worst-performing vehicles over time. If the standards are allowed to stagnate (i.e., remain the same for a significant period of time), they are not serving their primary purpose.

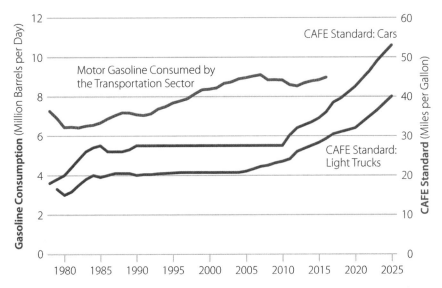

Figure 6-2. U.S. gasoline consumption increased while fuel economy standards stagnated. (Data from "Short-Term Energy Outlook, Oct. 2016," U.S. Energy Information Administration, 2016, Table 3.1, https://www.eia.gov/outlooks/steo/archives/oct16.pdf; United States, "2017–2025 Model Year Light-Duty Vehicle GHG Emissions and CAFE Standards: Supplemental," U.S. EPA 40 CFR Parts 85, 86, and 600; NHTSA 49 CFR Parts 531 and 533 §, 2011, http://www.nhtsa.gov /staticfiles.)

In the United States, after the 1973 oil crisis, the U.S. government enacted the first set of automobile fuel economy standards, which came into effect in 1978. From 1978 to 1985, the standards for passenger cars rose from 18 miles per gallon to 27.5 miles per gallon. In that time, U.S. motor gasoline consumption dropped from 7.3 million barrels per day to 6.7 million barrels per day, despite a growing U.S. economy and population.[8] Other factors probably influenced this drop as well, such as a change in the federal highway speed limit and higher gasoline prices. However, total passenger travel increased over this period, indicating that at least part of the reduction in motor gasoline demand was due to more efficient vehicles.[9]

Unfortunately, the standards were then allowed to stagnate for two decades. Despite various attempts, it was not until December 2007 that legislation tightening the standards was passed. Motor gasoline consumption rose considerably in the intervening decades, reaching 9.1 million barrels per day in 2007 (Figure 6-2). This stagnation of standards resulted in a tremendous loss to the

U.S. economy: 1–3 million barrels of oil per day,[10] worth tens of billions of dollars per year and hundreds of millions of tons of greenhouse gas emissions and other air pollutants that cause illness and premature death.

To avoid these negative outcomes, policies should have a built-in mechanism for tightening standards.

One way in which a standard might build in continuous improvement is to specify a formula for increasing stringency (3–6 percent per year is a reasonable guideline). This provides the greatest clarity and certainty to manufacturers, but it risks being unachievable if technical or physical bottlenecks are reached, which becomes increasingly likely after the standard has been tightened for many years, all low-hanging fruit is gone, and only the hardest technical challenges remain. On one hand, it may not be worth worrying about this when designing a standard, because the combined power of technological innovation and free markets continuously cough up huge advances, so today's efficiency frontier often becomes tomorrow's benchmark. Alternatively, it is possible to include a "safety valve," for example, allowing the standards to stop tightening if an independent technical review board determines that further improvements do not meet a cost-effectiveness test.

Another approach is to set the standard based on the most efficient vehicles already in the marketplace, plus a technology improvement factor, as is done in the Top Runner program in Japan.[11]

Focus Standards on Outcomes, Not Technologies

Vehicle performance standards should not specify the technology to be used but rather the physical outcomes they aim to achieve. For example, a good standard will regulate the amount of CO_2 emitted for each vehicle-kilometer (or when it is more applicable, each freight-ton-kilometer) traveled or transported. It should not specify, for example, that a particular engine design be used. This leaves companies the greatest latitude for innovation, so they can seek out the least expensive or most efficient means of achieving the standard.

Prevent Gaming via Simplicity and Avoiding Loopholes

Potentially the greatest difficulty in designing good vehicle performance standards is avoiding gaming and loopholes. For example, standards that vary based on vehicle weight (which allow greater fuel consumption for heavier vehicles) may encourage manufacturers to make heavier, less efficient vehicles.

In the United States, standards that were more lax for light trucks than for cars encouraged manufacturers to categorize more of their vehicles as sport utility vehicles (SUVs, technically a type of light truck) and to promote these larger vehicles to consumers.

Some principles that help to avoid loopholes include:

- Keep the standards simple. Write in clear and unambiguous language the quantitative performance outcome to be achieved. Carving out exceptions or special cases for various technologies or for vehicles with various properties opens up opportunities for gaming.
- Create real-world, in-use testing protocols, with real damages for cheats or nonperformance. Weak or inaccurate tests allow manufacturers to produce vehicles with nice fuel economy stickers but lousy real-world performance. Language that says "tests will be modified from time to time to better represent real-world conditions" and sets a fair horizon for future test requirements (e.g., 5 years) can push manufacturers to design for the world rather than the test.
- If the standard must vary by vehicle characteristics, it should vary by vehicle footprint, because this is more resistant to being artificially increased than a vehicle's weight or volume.
- If the standard must vary by vehicle characteristics, it should do so smoothly rather than ratcheting up or down suddenly when the vehicle characteristic crosses a certain threshold. For example, a vehicle performance standard might specify that the minimum required performance is equal to the vehicle's footprint multiplied by a coefficient rather than by breaking vehicles into distinct categories based on their footprints falling within particular ranges. A standard that changes abruptly makes a stairstep pattern when graphed against the characteristic on which the standard is based (such as vehicle footprint). Manufacturers will design vehicles that all fall at one edge of a stairstep, constraining their design choices and possibly increasing vehicle price without any emission benefit relative to setting a standard as a smooth line or curve that passes through the endpoint of each stairstep (Figure 6-3).
- Similar standards should apply to vehicle types that are substitutes for one another, such as passenger cars and SUVs (which are overwhelmingly used as on-road vehicles in the same functional capacity as passenger

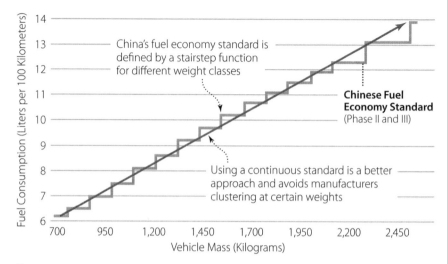

Figure 6-3. A continuous standard would avoid problems with China's stairstep fuel economy standard. (Drew Kodjak, "Global Trends in Passenger Vehicle Fuel Economy Standards," GFEI Fuel Symposium, Paris, 2014, https://www.slideshare.net/FIAFoundation/paris2014-drew-kodjak. Data reproduced with permission from the International Council on Clean Transportation under a Creative Commons Attribution-ShareAlike 3.0 Unported license.)

cars). This reduces the incentive for manufacturers to market and sell primarily the vehicle type with the weaker standards.

Additional Design Considerations
If vehicle performance standards are designed according to the aforementioned principles, there are only a few remaining points to watch out for.

Use Emission-Based Standards for Environmental Goals
If the policy goal is to reduce emissions, it is best if vehicle performance standards are based not on fuel consumption but on pollutant emissions per distance (or per unit cargo-distance) traveled. The effect is largely similar, but there are two key differences:

- For local air pollutants, an emission-based standard allows control technologies, such as particulate filters, to be used to help meet the standard. This helps ensure the standards are technology finding (rather than restricting manufacturers' options to choose among technologies addressing fuel efficiency), allowing compliance at least cost.

- For greenhouse gases, an emission-based standard should include leakage of refrigerant from the vehicle's air conditioning system. These refrigerants can be powerful greenhouse gases, and the fuel efficiency of the car has no relationship to their leakage rate.

On the other hand, if the policy goal is to reduce petroleum consumption, then vehicle fuel consumption performance standards are more appropriate.

Build Industry Support

Vehicle manufacturers may either oppose the enactment of tighter vehicle performance standards (as they did in the United States in 1990) or work constructively with policymakers on a compromise (as they did in the United States in 2009).[12] The support of automakers can make a big difference in the likelihood that a legislature or a regulatory agency feels comfortable enacting suitable performance standards, and it may avoid the uncertainty and delay associated with lawsuits challenging the regulations.

The core elements of smart policy design will help garner the support of automakers who have serious commitments to new technology. In particular, automakers value reasonably long lead times, regulatory certainty, flexibility in how a standard is met (technology-finding policy), and the opportunity to trade, bank, or borrow credits. Of course, companies that fail to make meaningful technology development investments may still oppose tighter standards, but it makes little sense to design policy to protect the worst performers.

Industry players, especially vehicle manufacturers (as opposed to vehicle component manufacturers), are more likely to be supportive if they are given a place at the table early, a voice in setting the starting stringency of the new standards, and regulatory certainty through a known schedule of improvements extending many years into the future. Attempts by industry to set up periodic review of the standards should be viewed with skepticism, as reviews add to uncertainty and are often used to undermine the stringency of existing standards. If a review is called for, it should be conducted by qualified, independent experts.

Some vehicle and engine manufacturers may be supportive if they believe R&D is among their competitive strengths. Cummins is a U.S. company that manufactures engines and power generation equipment. John Wall, Cummins's chief technology officer, explained why Cummins has supported increased vehicle performance standards in the past: If Cummins knows the standards it will need to meet, it can invest in developing the necessary technologies, such as hybrid power trains or heat recovery systems. When the standard comes into effect, Cummins can provide standard-compliant products that are better and cheaper than those of their competitors, and Cummins gains market share and a return on its R&D investment.[13]

Use Test Procedures Resembling Real-World Driving Conditions and Retest Randomly Selected Vehicles

Vehicle performance standards can be undermined by test procedures that are intentionally or unintentionally designed to allow manufacturers to achieve good test results while achieving worse real-world, on-road performance. This has been a particular problem in Europe, where the gap between vehicles' tested performance and their actual performance continues to grow, as manufacturers game the test procedure: This gap was 10 percent in 2001 and grew to 35 percent by 2014.[14] This means the average car in Europe that appears to comply with the standard actually burns 35 percent more fuel than demonstrated during the test procedure. Vehicle manufacturers are able to supply specially prepared vehicles for testing; for example, they can take out the seats to make the car lighter, tape the seams to make it more aerodynamic, and fill the tires with liquid to reduce their rolling resistance. They can even select an especially favorable test track to be used.[15] This "golden car" approach is gaming the system across the board, but because of poor policy design, it is legal.

Vehicles should be tested upon introduction to market and retested during the vehicles' lifetime. Testing should be carried out by an independent third party and on randomly selected vehicles, not vehicles specially prepared for testing by manufacturers. Laboratory tests should be supplemented by on-road tests that reflect actual driving conditions. Some car and part manufacturers and international bodies are pushing for internationally harmonized standards for vehicle testing procedures, which would arguably strengthen enforcement and better identify underperformers. However, some nations and regions (such as the United States and the European Union) prefer to have their own testing procedures to better account for local conditions. The success of internationally harmonized testing standards will depend on the ability of stakeholders to implement standards that promote rigorous, randomized testing in accordance with the principles outlined above.

Case Studies

Japan's Top Runner Program for Passenger Vehicles

Japan has had vehicle performance standards for passenger vehicles since 1979, but in 1999 they were brought under the scope of Japan's new Top Runner program.[16] The program first identifies the most fuel-efficient vehicle in each weight class every few years. The efficiency standard is then updated to reflect the efficiency of this "top runner" vehicle but adjusted to account for potential technological improvement over the next set of years the standard covers.[17] This technological improvement percentage, which is determined by regulators and subject to a public comment period, pushes manufacturers to achieve vehicle efficiency above and beyond what is currently in the market, incorporating the principle of continuous improvement discussed earlier.[18] To push the boundaries of technology, tax breaks are offered to companies that meet the performance standards years ahead of schedule.[19]

The Top Runner program sets standards based on the sales-weighted average efficiency of vehicles sold by each manufacturer,[20] so manufacturers can sell some vehicles that do not meet the standard as long as they sell enough vehicles that exceed the standard by a sufficient margin. The Top Runner program also stipulates "display items," or statistics about each product that must be displayed in places that can be readily seen by buyers, such as in catalogs and on exhibits. The data to be included vary by product. For passenger vehicles, required data include the engine type, vehicle weight, riding capacity, energy consumption efficiency, emissions, and other details.[21]

Between 1999 and 2010, the Top Runner program improved the energy efficiency of passenger vehicles by 23 percent and improved the efficiency of small freight vehicles by 13 percent.[22]

Although the Top Runner program has seen some success, it has significant drawbacks. One limitation is that the program categorizes vehicles by weight class rather than by footprint, encouraging manufacturers to build bigger vehicles. Another limitation is that although the standard includes a mechanism for increasing stringency, this mechanism relies on technology that is already in production in the baseline year, resulting in standards that are easily met. For example, although the program has resulted in a 49 percent improvement in efficiency, this improvement has taken place over 15 years, delivering only a 2.7 percent improvement per year. Third, Top Runner uses a stairstep function in setting efficiency requirements and therefore encourages gaming. It's worth noting that this type of program is also much easier to implement once top-performing technologies are already available; it would not be as effective for an initial program in countries that are new to performance standards. More detail on economic incentives to purchase more efficient cars is provided in Chapter 7 ("Vehicle and Fuel Fees and Feebates").

China's Vehicle Fuel Consumption Control Program

China has introduced vehicle fuel economy standards in four phases, with the first phase covering 2004–2007, the second phase 2008–2011, the third phase 2012–2015, and the fourth phase taking effect at the start of 2016.[23] Although the program currently fails to incorporate many of the design principles discussed earlier, policymakers are taking lessons from global leaders in vehicle performance standards to improve on several of its current deficiencies. China's standards have seen some success and hold some promise for future efficiency improvement.

China's vehicle fuel consumption standard varies based on vehicle weight, not footprint, and establishes different standards for different technologies (normal, SUV, and minivan). Each phase extends only about 4 years, failing to give businesses certainty and a long-term planning horizon. New standards are rolled out only 3 years before manufacturers need to demonstrate compliance, which results in slower and more expensive technology upgrades, as manufacturers are limited in their ability to plan for the long term. The standard is implemented as a series of stairsteps rather than a continuous function (although in the future it will probably move to a continuous function) and relies heavily on manufacturer input. Lastly, policies focus on fuel consumption rather than pollutant emissions.

For compliance and enforcement, China uses a European test protocol for monitoring, verification, and enforcement, but Europe's test protocols (as discussed earlier) allow extensive gaming by manufacturers. Compliance is also demonstrated using preproduction and production models tested in labs rather than testing models on the road. Furthermore, there are no compliance penalties and no compliance follow-ups once vehicles are on the road. China is creating its own testing protocol, but whether that will turn out to be tougher or looser than the European approach remains to be seen.

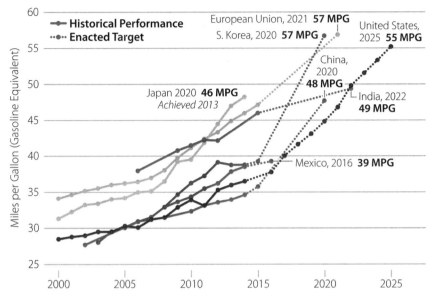

Figure 6-4. The strongest fuel economy standards in the world. ("Passenger Car Miles per Gallon, Normalized to CAFE," 2015, http://www.theicct.org/sites/default/files/info-tools /pvstds/chartlibrary/CAFE_mpg_cars_Sept2015.pdf. Data reproduced with permission from the International Council on Clean Transportation under a Creative Commons Attribution-ShareAlike 3.0 Unported license.)

Taken together, the Phase I and Phase II standards reduced real-world average fuel consumption of vehicles by approximately 12 percent, a reduction that would have been greater but for the fact that the standards encouraged the production of higher-fuel-consuming cars with larger engine displacement, eliminating some of the reduction in total fuel consumption.[24]

On the upside, the efficiency requirement of China's standard is aggressive, making the Chinese fuel economy standard one of the most stringent in the world (Figure 6-4).[25] China supplements its fuel economy standards with a variety of other policies, including fuel taxes; subsidies for fuel-efficient, plug-in hybrid, and all-electric vehicles handled upstream of the consumer; a commitment to scrapping old and high-emitting vehicles (although the policy mechanism has not yet been announced); and a tax and rebate that together encourage the purchase of smaller engines.[26] China also requires labels disclosing the fuel efficiency of vehicles, although as of 2012, one study found only a 62 percent compliance rate with requirements for cars to display these labels.[27]

The Chinese fuel economy standards are for the most part well designed and effective, although they would benefit from better testing and enforcement procedures and from being based on footprint rather than weight.

The United States' Corporate Average Fuel Economy Standard

The United States' fuel economy policy is the Corporate Average Fuel Economy (CAFE) standard. First coming into force in 1978, the CAFE standard aims to improve the average fuel economy of newly sold light-duty vehicles. CAFE is applied to a manufacturer's current model-year fleet of passenger cars or light trucks, manufactured in the United States and with a gross vehicle weight rating of 8,500 pounds (3,856 kilograms) or less. Manufacturers are required to meet an average fuel economy target based on the sales-weighted mean miles per gallon in a given year.

The penalty for noncompliance is steep: If a manufacturer's new vehicle fleet falls below the applicable standard, the manufacturer must pay a penalty, currently U.S. $55 per mile per gallon below the standard, per vehicle manufactured for the U.S. market.

In 2012, after passage of the 2007 Energy Independence and Security Act, the National Highway Traffic Safety Administration established a credit trading mechanism to allow manufacturers to transfer compliance credits between car and truck categories, carry compliance credits forward up to five model-years, and transfer compliance credits between firms. During this revision, the method of calculating compliance was changed to the product of the vehicle's wheelbase and its average track width (footprint). On one hand, the change loosened the standards by allowing larger vehicles to meet less aggressive fuel economy requirements than smaller vehicles, reducing the incentive to sell more small cars. On the other hand, by indexing the standard to vehicle size rather than weight, this formula builds in a crucial incentive to reduce weight, which enables improved fuel economy and can improve overall roadway safety.

The U.S. CAFE standard is often seen as a success and in recent years has significantly driven fuel economy improvements. However, without a mechanism to regularly improve the stringency of the standard, fuel economy stagnated for nearly 20 years between 1990 and 2010. Given the current administration in the United States, it is unclear whether the standards will continue to strengthen in the future or whether they will stagnate again.

In addition, the standards are set separately for passenger cars and light-duty trucks, which has encouraged manufacturers to make small modifications that reclassify passenger cars as light-duty trucks, thereby lowering the minimum required fuel economy.

Another issue is the inclusion of a review process for future year standards. During the 2012 extension of the standards, policymakers included the option of a midterm review of the standards, which manufacturers are trying to use as a mechanism for reducing the standards stringency.

Nevertheless, the long time horizon, flexible compliance options, and use of standards based on footprint (rather than weight) have helped the U.S. CAFE standards improve vehicle fuel economy in recent years.

Conclusion

Vehicle performance standards are one of the best policies for reducing emissions in the transportation sector. When properly implemented, they drive down emissions year after year while achieving net savings for society. The best standards in the world are publicly known many years in advance, have built-in mechanisms to tighten the standards over time, are technology finding, and are resistant to gaming. Well-designed standards and proper enforcement will be a key part of the transition to a clean energy future.

CHAPTER SEVEN

Vehicle and Fuel Fees and Feebates

Along with performance standards, fees on fuel and inefficient new vehicles are among the best policies for reducing emissions from on-road vehicles, which make up 71 percent of emissions from the global transportation sector.[1] Both fuel and vehicle fees have been widely used in the past, generating revenues for infrastructure projects such as road construction and public transit. With careful government investment of these revenues—targeting improvements such as urban mobility measures, efficiency, and research and development (R&D) for new clean energy and efficiency technologies—it is possible to magnify their impact and accelerate the transition to a zero-carbon future. Vehicle and fuel fees and feebates can provide about 1 percent of the reductions required in a two-degree warming scenario (Figure 7-1).

Policy Description and Goal

Fuel fees are taxes levied on vehicle fuels. They should reflect all negative externalities caused by fuels and should be implemented upstream in the supply chain. Revenues from fuel fees can be used to mitigate regressive impacts or to accelerate air pollution or greenhouse gas reduction programs.

Vehicle feebates provide a rebate to purchasers of efficient vehicles, funded by fees levied on purchasers of inefficient vehicles. For feebates to be effective, governments can reduce regulatory hurdles and minimize the effort involved in obtaining a rebate. Feebates should apply to all vehicles in the same class (e.g., all passenger light-duty vehicles).

Fuel fees and vehicle feebates can encourage increased new vehicle fuel economy while reducing vehicle distance traveled. Fees and feebates should be long-term policies with a clear implementation schedule. Their stringency (set points and strength) can be adjusted over time as technology improves. They can contribute at least 1 percent of cumulative emissions reduction to meet a two-degree target through 2050.

This chapter will cover three related types of policy:

- Fees charged on carbon-based fuels
- Fees charged on newly sold on-road vehicles, primarily passenger cars, SUVs, and freight trucks
- A feebate, a fee or a rebate for newly sold vehicles varying based on their efficiencies

Figure 7-1. Potential emission reductions from feebates. (Analysis done using data with permission from the International Institute for Applied Systems Analysis [IIASA]. Data source: Tavoni et al., 2013. Data downloaded from the LIMITS Scenario database hosted at IIASA, https://tntcat.iiasa.ac.at/LIMITSPUBLICDB/dsd?Action=htmlpage&page=about.)

Vehicle fees and feebates are most effective in helping reduce fuel consumption of on-road vehicles. Commercial airliners, large cargo ships, and trains have decades-long lifetimes and high manufacturing costs, and therefore a fee or feebate of sufficient magnitude to substantially affect the economics of buying and operating one of these nonroad vehicles over its lifetime may be politically difficult.

Fuel fees, on the other hand, have an ongoing effect on the economics of aircraft, rail, and ships over these vehicles' lifetimes and may be more effective at influencing manufacturers to improve nonroad vehicle efficiency. Significant improvements are possible; for instance, new aircraft designs could cut fuel consumption by up to 40 percent.[2] However, because fuel and vehicle fees and feebates apply mostly to on-road vehicles today, the remainder of this chapter will consider only on-road transportation modes.

Fees on Carbon-Based Fuels

In some countries, fuel taxes are implemented at multiple levels: The national government, states or provinces, and localities can all charge taxes on fuels. In most cases, the fee is implemented as an excise tax, or a tax on a particular volume of fuel sold. (In contrast, a sales tax is based on the purchase price.) Some U.S. states have taxes that vary with the underlying price of fuel or other factors.[3] Fuel taxes are levied on the seller, who often passes some or all of the cost increase on to consumers.

Fuel taxes influence drivers via two different mechanisms: how much to drive existing vehicles and which new vehicles to purchase. Fuel taxes increase travel costs, which will cause some people or businesses to reduce travel demands or shift trips to a different form of transportation (such as biking, walking, or public transit for people; or for freight, a more efficient mode per ton of goods transported, such as rail). This effect can happen quickly, because it does not require vehicle fleet turnover.

When a consumer or business is ready to buy a new vehicle, they may consider the cost of fuel when deciding which model to purchase. More fuel-efficient models will be more attractive if fuel prices are higher. Because this effect depends on the vehicle fleet turnover rate, fuel prices must be sustained at a higher level for an extended period (if not permanently) for this effect to significantly improve fleetwide fuel economy.

Fees on carbon-based fuels are particularly useful to counteract the rebound effect of improved vehicle efficiency. If vehicles become more efficient (perhaps because of vehicle performance standards or market demand for more efficient technologies), the cost of driving goes down. This induces people to drive more, which offsets some emission benefits of vehicle efficiency (about 10 percent for passenger vehicles[4] and 15 percent for trucks[5] in the United States). Fuel taxes can offset this rebound effect.

It is important to note that fuel fees are not likely to result in significant behavioral changes, at least at politically feasible levels. Therefore, fuel fees are particularly valuable as a revenue raising mechanism. Revenue from fuel fees can be used for other policies or programs, such as clean vehicle rebates or public transit infrastructure, which can further reduce emissions and provide the public with transportation alternatives.

Fees on Newly Sold Vehicles

In addition to applying fees to transportation fuels, some regions also charge a fee when new vehicles are purchased. This may come in the form of an excise tax or a permit to own and operate the vehicle in a given area. For example, the Shanghai region auctions permits for new vehicles, and permit prices can be considerably more expensive than the vehicle purchase price itself.[6]

Although a flat fee per vehicle can help encourage shifting to other transportation modes, it does not encourage purchasing more efficient cars over

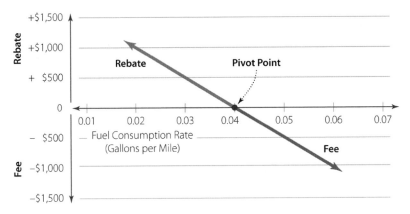

Figure 7-2. Illustrative design of a feebate.

less efficient ones. However, a fee can vary based on car efficiency or fuel type to achieve this goal. For example, in Norway, high auto taxes are waived for battery electric vehicles, although these incentives have started to be phased out.[7] Similarly, Denmark formerly waived its high taxes on new vehicle sales on electric cars, although a new, center-right government began phasing out this exemption,[8] causing electric vehicle sales to drop 60 percent in 1 year.[9] This illustrates the importance and effectiveness of these policies in affecting consumer buying decisions.

Feebates

A feebate is a fee combined with a rebate under one policy. Buyers of inefficient cars are charged a fee at the time of sale, and buyers of efficient cars receive a rebate. Regulators determine a policy target, referred to as a pivot point, as the efficiency level at which a car neither incurs a fee nor receives a rebate (Figure 7-2). The feebate rate (how quickly the fee and the rebate escalate as one moves away from the pivot point) and whether it is levied as a tax or rebate is then determined based on how much a new vehicle's fuel efficiency is better or worse than the target.

The government may design the feebate with its revenue goals in mind. For instance, setting a higher efficiency target can make more vehicles pay the fee than will receive rebates, resulting in a revenue stream to fund public transportation or other government projects.

Conversely, setting the target at a level where the sum total of all fees in a year equals the sum total of all rebates paid can make the feebate revenue neutral. In order to maintain revenue neutrality, the target must be adjusted frequently to keep up with changes in the efficiencies of cars offered by manufacturers and the buying preferences of the public.

It is also possible to reward extremely efficient cars (those with fuel efficiency far better than cars at the target) by establishing an efficiency level beyond which the feebate offers an increased subsidy rate. Offering a higher subsidy rate for extremely efficient vehicles can help promote the development of more advanced efficiency technology.

When to Apply This Policy

Fees on Carbon-Based Fuels

Fees on carbon-based fuels can achieve three central goals of transportation vehicle policy: They discourage driving by making fuel more costly; they encourage the purchase of new vehicles because older, less efficient vehicles are more expensive to drive; and they encourage consumers to purchase more efficient vehicles (which also encourages manufacturers to produce more efficient vehicles) because they cost less to drive. Therefore, fees on carbon-based fuels are broadly applicable.

Fuel fees also help incorporate the social costs of fuel consumption, such as the contribution to climate change or health impacts, into the price paid by consumers. Additionally, revenues from fuel fees are likely to be substantial, and they can be reinvested in projects aimed at reducing the impacts of fuel burning (e.g., by decarbonizing the economy and providing medical services). For these reasons, fuel fees are an appropriate and effective policy.

Vehicle Fees and Feebates

Vehicle fees are useful to reduce demand for travel by car and, if the fees vary with fuel efficiency, help steer consumers toward purchasing more efficient vehicle models and manufacturers toward designing more efficient models. Because consumers sometimes strongly discount fuel savings when deciding which car to buy,[10] fees and feebates can improve the efficiency of newly purchased vehicles by moving some of the economic impact upfront, where it may be considered more thoroughly by vehicle purchasers. A small fee on conventional fuel vehicles could also be used to create a sustainable budget for subsidizing electric vehicles and other alternative fuel vehicles.

Feebates are useful for incentivizing the development and deployment of new, more efficient vehicle models. By offering a payout that scales with efficiency, feebates reward buying the most cutting-edge, efficient vehicles. Manufacturers are incentivized to conduct R&D and manufacture more efficient models in order to take advantage of this rebate and sell more cars. This means feebates are a particularly good complement to vehicle performance standards, which are especially effective at improving the worst- or lowest-performing vehicles.

A flat vehicle fee, such as an annual registration fee, can be helpful for congested cities, such as Shanghai, where it limits the number of cars on the roads and generates substantial revenue for improving the attractiveness and convenience of other transportation modes such as walking, biking, and public transit. In contrast, a graduated fee or a feebate is best when the main goal is incentivizing the development and sale of more efficient vehicles, and it works better at a larger scale (such as national policy) than at the local or even the state level.

A feebate can lead to odd outcomes if it applies only to a specific region, because people who want to buy inefficient vehicles can purchase them from a neighboring region without a feebate. Meanwhile, people who want to buy efficient vehicles purchase them in the region where the feebate will benefit them. This phenomenon has actually happened in Shanghai, where residents purchase cars in neighboring provinces to avoid Shanghai's expensive license plate fee. As a result, the government pays for many rebates while taking in little money in fees, thus not achieving the policy outcome of increasing the share of efficient vehicle purchases. Charging a fee upon vehicle registration rather than at the point of sale could help mitigate this loophole and ensure the feebate is more accurately enforced.

Detailed Design Recommendations

Policy Design Principles
The following policy design principles apply to vehicle and fuel fees and feebates.

Create a Long-Term Goal and Provide Business and Consumer Certainty
Fuel taxes should be levied indefinitely, because fossil fuel externalities do not diminish with time. Long-term fuel tax certainty empowers manufacturers to invest in R&D projects improving fuel efficiency, knowing there will be a

market for these vehicles years down the road when they are ready. It also gives consumers a stronger incentive to buy more efficient vehicles. Consumers uncertain whether a fuel tax will be maintained in the future might buy a less efficient car in hopes the tax will be reduced or suspended.

Taxes and fees should always be indexed to inflation so they remain consistent in real terms. The tax may need to increase if vehicles become more efficient, to prevent the incentive from being eroded and to prevent an increase in driving caused by the rebound effect (discussed earlier). One option is to build in an adjustment mechanism that indexes the tax to the average vehicle fleet efficiency.

Vehicle fees and feebates should similarly be publicly known many years in advance to maximize their incentive effect and ensure that consumers do not hold off on purchasing a new vehicle in hopes of getting one when the fee lapses. Vehicle fees and feebates may not be permanent—for example, if the transportation fleet transitions to clean energy and if congestion in urban areas is no longer a concern—but because these goals are many years away, vehicle fees and feebates are likely to remain valuable policy tools for the foreseeable future.[11]

Price In the Full Value of All Negative Externalities for Each Technology or Use a Price-Finding Mechanism

Fuel taxes should price in the full value of all social harms. These include public health impacts due to local pollutant emissions, climate change impacts, congestion impacts, infrastructure impacts, and traffic accidents. Fuel taxes could also be applied on an energy-equivalent basis, rather than deriving the externality costs for each fuel, which could be difficult in some circumstances.[12]

Some of these negative impacts, such as traffic accidents, apply to all vehicles regardless of what kind of fuel they use. In these cases, a separate fee could be levied on all vehicles to cover those impacts, in which case it need not also be factored into a fuel tax.

A price-finding mechanism is most appropriate for vehicle fees in crowded urban areas where the maximum number of cars is known. Auctioning permits is a straightforward way to accomplish this.

Adjustment of the feebate pivot point is also price finding, because the adjustment accomplishes a specific revenue-related goal (e.g., achieving revenue neutrality or obtaining a particular amount of net revenue for other projects). The correct pivot point price is revealed by buyers' choices.

The rate of a feebate might also be designed to be price finding. The policy-maker first needs to identify a desired performance outcome—in this case, a specific magnitude shift in the efficiency of the average car sold. Then, various feebate rates can be tested, either in the real world or through studies, to find the feebate rate that achieves the targeted efficiency increase.

In some instances, particularly for vehicle and fuel fees, the additional charge may not result in large behavioral changes that eliminate the externalities. Therefore, these policies are particularly helpful as a revenue mechanism to create funding for rebates and other programs to promote the adoption of more efficient, cleaner vehicles and to pay for alternatives to car travel, such as public transit.

Eliminate Unnecessary Soft Costs

If significant regulatory hurdles to the purchase or operation of a vehicle exist in a given region, those hurdles could be lowered for particularly efficient or zero-emission vehicles, making it faster and easier to obtain these models.

For example, electric vehicles could receive expedited permit processing or a minimum number of permits in regions that conduct auctions for vehicle permits.

If a rebate is offered on efficient or zero-emission vehicles, the rebate can be handled by the dealer (or the manufacturer, in the case of manufacturers who sell directly to the public). That way, the dealer or manufacturer is responsible for submitting relevant government paperwork and simply includes the rebate value in the car price seen by consumers. However, careful monitoring is needed to ensure manufacturers do not game the feebate system under this approach by applying for the rebates and retaining the subsidy.

Capture 100 Percent of the Market and Go Upstream or to a Pinch Point When Possible

It should not be easy for manufacturers to substitute other types of fuel or vehicles that have similar negative externalities simply to evade fuel or vehicle fees.

If a fee is applied only to petroleum gasoline, and cars are capable of burning ethanol, this risks consumers switching to ethanol to avoid the fee. Therefore, fuel fees should be set on *all* carbon-based fuels, not solely on petroleum fuels or on gasoline. Fees should be based on pollution emission intensity (grams of each pollutant emitted per unit usable energy in each fuel) or carbon intensity of fuel so consumers are still encouraged to purchase the lowest carbon-emitting fuel type.

One of the main hazards to avoid is applying different fees, or a feebate with different pivot points, to vehicles distinguished by characteristics readily modified by manufacturers. For example, automakers started manufacturing and marketing SUVs to consumers partly because they were subject to weaker vehicle performance standards than were passenger cars.[13] This led consumers to buy larger and less efficient SUVs, impairing the policy effect.

Ensure Economic Incentives Are Liquid
Rebates associated with feebates should be provided as an immediate discount on the vehicle's purchase price or a promptly delivered cash payment.

If the payment is issued as a tax credit, even a refundable tax credit, the psychological impact of the rebate on consumers is diluted by separating it in time from the vehicle purchase. It also increases hassle by requiring consumers to take extra steps (on their tax returns) to receive it.

Additional Design Considerations
If vehicle and fuel fees are designed according to the principles described earlier, only a few remaining points must be tackled.

Mitigate Regressive Impacts of Fees
Like sales taxes, fuel and vehicle fees tend to be regressive; they disproportionately affect people with lower income. People with lower incomes spend a higher percentage of their income on transportation, and thus a higher percentage of their income goes into these fees.

Additionally, lower-income groups may have older and less efficient vehicles, causing them to buy more fuel per mile traveled, increasing the amount of fuel-based fees they pay.

Regressive fees can be mitigated through socially aware use of the resulting revenues. Revenues funding better public transit, particularly urban bus and metro systems serving low-income neighborhoods, provide alternative means of mobility to low-income residents at affordable prices. Funds can also be directed to nontransportation programs benefiting low-income residents, such as improving the quality of schools in low-income areas.

Another option is to rebate the fees to society as a flat payment to each person (a "dividend") or a graduated dividend with higher values going to lower-income people. (Dividend payments should not be based on fuel use or fees paid in order to avoid eroding the incentive created by the fees.) With smart use of funds, the benefits to low-income residents can significantly outweigh added costs from fuel and vehicle fees.

Build Industry Support
Fuel and vehicle fees will be easier to enact, and face fewer legal challenges after enactment, if automakers do not oppose these policies. Although vehicle manufacturers are likely to oppose straightforward vehicle fees, they might not oppose a feebate if they believe rebates on more efficient cars can allow them to increase sales by a greater margin than the inefficient car fees will cost them. Right now the opposite tends to be the case: SUVs are the high profit-margin vehicles for manufacturers, and electric vehicles bring in much less profit.

Policymakers should discuss feebate design, such as the feebate rate and the pivot point, with the vehicle industry, attempting to assuage industry concerns. Placing the pivot point so the policy is slightly less than revenue neutral (meaning the feebate is a slight net expenditure) is one way to ensure that vehicle manufacturers, in the aggregate, benefit from the feebate.

Vehicle manufacturers might oppose fuel fees unless they are convinced they will not appreciably reduce demand for their products. Electric vehicle manufacturers, or other vehicles whose fuels are not subject to the fee, might see increased sales of those vehicle models.

Similarly, manufacturers that offer particularly efficient petroleum-powered car models might expect to gain market share from competitors and thus benefit from fuel fees. Manufacturers whose offerings consist of vehicle models that burn carbon-based fuels and are less efficient than competitors' models are likely to be the hardest to convince to support fuel fees.

Case Studies

Federal Gasoline Taxes in the United States
The U.S. federal gasoline tax has achieved only a fraction of its potential because of poor policy design. The United States first imposed a gasoline tax of 1 cent per gallon in 1932. It has periodically been adjusted since then, most recently in 1993.[14] The tax is not indexed to inflation, so over time the true tax value decays (the "real rate"), whereas the tax face value remains constant (the "nominal rate"). However, the tax burden on drivers is also affected by the

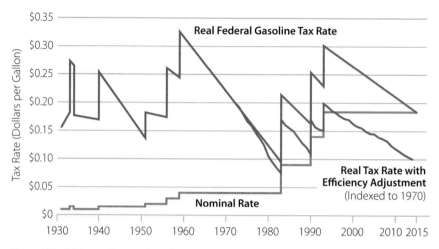

Figure 7-3. Without adjusting for inflation, the U.S. gasoline tax has continuously lost value. (Image created from written description in Kevin McCormally, "A Brief History of the Federal Gasoline Tax," Kiplinger, 2014, http://www.kiplinger.com/article/spending/T063-C000-S001-a -brief-history-of-the-federal-gasoline-tax.html.)

efficiency of the vehicles they drive; vehicles that go farther per gallon of gasoline result in a lower tax bill per mile for drivers. Figure 7-3 shows the nominal rate, the real rate, and the real rate after adjustment for improving efficiency of U.S. cars since 1970 (indexed to 1970).

In real terms, the tax rate is slightly below historical average levels. However, as cars become more fuel efficient, the gasoline tax generates comparatively less revenue than it once would have at the same level. Because gasoline tax revenue is often used to fund infrastructure, the net effect has been less government funding available for road infrastructure construction and maintenance.

Before the interstate highway system, gas tax revenue was devoted to deficit reduction and war spending. In 1956, the Highway Trust Fund was established to help pay for the new interstate highway system, and all gas taxes were devoted to this fund.[15]

For a period in the 1990s, some of the tax revenue was again directed to deficit reduction, but this ended in 1997, because the tax was insufficient to maintain the solvency of the Highway Trust Fund. Since then, the gas tax has proven insufficient, and the Highway Trust Fund has lost money each year. The shortfall will reach $80 billion by 2018.[16]

One important problem with the U.S. federal gasoline tax is the failure to index it to inflation. If the tax had been indexed to inflation starting in the 1930s, no legislative action would have been necessary over the ensuing decades to maintain tax viability, and periods when the tax was allowed to decay for many years (such as 1959–1983 and 1993 to the present day) would have been avoided. This would also have helped ensure the continued solvency of the Highway Trust Fund.

The other major problem with the U.S. gasoline tax is that it does not capture the full value of social harms from driving a gasoline-powered vehicle. Although under the Highway Trust Fund the tax was designed as a means of generating revenue to maintain transportation infrastructure, not account for external costs to society, gas taxes should ultimately rise to the level of social costs generated by driving. To capture all social costs, the gasoline tax, alongside a fuel use tax, would together need to reflect:

- Climate change damages
- Premature deaths and illnesses from localized air pollutants
- Traffic accidents
- Congestion, lost time, and reduced productivity
- Subsidies and tax breaks received by the oil industry (because these are funded by taxpayers at large, a harm to society)
- Indirect subsidies received by drivers, such as free parking (the cost of which is borne by the government, at least on public roads and in public lots, which are taxpayer-funded)
- Military spending for purposes of protecting oil supplies

Government should strive to internalize the value of all externalities to the extent it is practical.

France's Bonus-Malus Feebate Program

The largest automobile feebate program in the world is the French Bonus-Malus program.[17] The program entered into force in January 2008 with three goals: steering buyers toward vehicles that emit less CO_2, encouraging the development of new low-emission vehicle technologies, and accelerating retirements of old, inefficient vehicles. The pivot point of the feebate is automatically revised downward (requiring vehicles to be more efficient to avoid the fee) every 2 years, maintaining revenue neutrality.

Like most feebates, the Bonus-Malus program provides a rebate for efficient

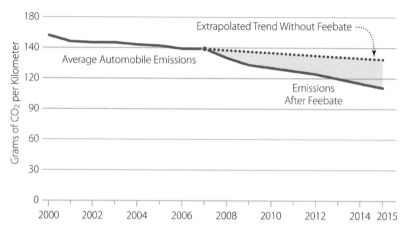

Figure 7-4. France's feebate drove an improvement in the vehicle emission rate. ("Évolution du Marché, Caractéristiques Environnementales et Techniques des Véhicules Particuliers Neufs Vendus en France," Agence de l'Environnement et de la Maîtrise de l'Énergie, September 2017, http://www.ademe.fr/sites/default/files/assets/documents/evolution-marche-vehicules -neufs-2017-8524.pdf. Historical fuel economy data in numerical form courtesy of Wikimedia under Creative Commons Attribution Share-Alike license.)

car buyers, maxing out at €6,300 for the most efficient vehicles, and a fee of up to €8,000 on the least efficient vehicles (as of 2016). The bonus cannot exceed 27 percent of the vehicle's cost, and diesel vehicles are not eligible for a bonus.[18] Unlike a traditional feebate, an annual penalty of €160 is assessed on owners of high-emitting vehicles, helping accomplish the goal of accelerating retirements.[19] One downside to the program is its stairstep function, rather than continuous function, which has only slightly improved efficiency. Manufacturers responded by focusing on vehicles near the step points, allowing them to only incrementally increase efficiency while significantly increasing rebates.

The Bonus-Malus program was successful in accelerating the rate of efficiency gains in the French vehicle fleet, as shown in Figure 7-4. Along with other regional fuel efficiency policies, France's feebate contributed to a reduction of roughly 25 percent in CO_2 emissions per kilometer through 2015.[20]

Conclusion

Fuel fees and vehicle feebates are powerful policies for decreasing emissions by reducing driving and by encouraging more efficient vehicles. Fuel fees should reflect the social harms of burning fuel. These fees reduce fuel use and provide a source of revenue to help fund urban mobility projects. A feebate is a powerful incentive to buyers and vehicle manufacturers to opt for more efficient vehicles, and the pivot point can be adjusted to achieve the government's revenue goals. Together, fuel fees and vehicle feebates can help drive decarbonization of the transportation sector and usher in a clean energy future.

CHAPTER EIGHT

Electric Vehicle Policies

Today, mobility of goods and people is powered predominantly by oil. Most cars and motorbikes burn gasoline; most trucks and buses burn diesel fuel; and trains, ships, and aircraft typically burn other petroleum-derived fuels. To lower greenhouse gas emissions, it is crucial to deploy technologies that reduce emissions from the transportation sector—and from on-road vehicles in particular—quickly and at large scale. For example, Figure 8-1 shows the share of fuel use by vehicle type in the United States, with cars, light trucks, buses, and trucks making up on-road vehicles.

One promising technology to achieve decarbonization in the transportation sector is the electrification of on-road vehicles. Vehicle electrification policies can contribute at least 1 percent of cumulative emission reductions to meet a two-degree target through 2050 (Figure 8-2).

Policy Description and Goal

Electrification of on-road vehicles is an important part of decarbonizing the transportation sector. A number of policies can be used to encourage suppliers and accelerate consumer adoption, including rebates and subsidies, development of charging infrastructure, electric vehicle (EV) sales mandates, and consumer education. It is crucial that these policies be designed with a long time horizon, that subsidy rates keep up with technological progress, and that they be phased out gradually according to a schedule or formula set in advance. Programs in the U.S. state of Georgia and in China show the potential for policy to achieve success, as well as specific pitfalls to avoid. Good policy can hasten the day when EVs satisfy most on-road passenger transportation needs in cities around the world.

EVs offer two key benefits that help reduce emissions. First, EVs are three times as efficient as gasoline vehicles: 59–62 percent of the electrical energy is converted into power to turn the wheels, whereas a gasoline vehicle converts only 17–21 percent of the chemical energy in the fuel into useful work.[1] Second, it is possible for electricity to be generated using zero-emission technologies, such as solar panels, wind turbines, hydro dams, or nuclear power plants, meaning the operation of an EV can have close to zero emissions.[2]

EVs also offer economic benefits to their owners. Their efficiency means that they cost little to operate: A typical electric passenger car can travel 43 miles for $1 worth of electricity.[3,4] This is about one-fourth of the fuel cost of a typical 2016 gasoline-powered passenger car.[5] Additionally, EVs have far fewer moving parts than vehicles with internal combustion engines (they typically need no radiator or transmission), so they are more reliable and need less maintenance.[6]

If EVs have so many benefits, why do we need policy to help promote their commercialization and deployment? There are two major barriers slowing EV deployment. First, although their costs are falling rapidly, EVs still cost more than similar gasoline or diesel vehicles. Second, EVs need sufficient access to charging infrastructure. Most EV owners do most of their charging at home, but this may be a challenge for households that lack access to electricity in a garage or off-street

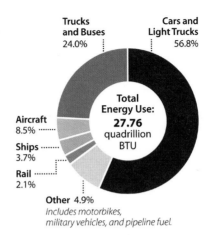

Figure 8-1. U.S. transportation sector energy use by mode, 2015. (U.S. Energy Information Administration, "Annual Energy Outlook 2017," n.d., https://www.eia.gov/outlooks/aeo/supplement/excel/suptab_36.xlsx.)

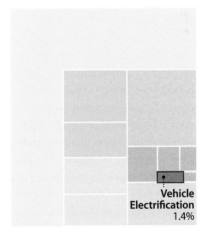

Figure 8-2. Potential emission reductions from vehicle electrification. (Analysis done using data with permission from the International Institute for Applied Systems Analysis [IIASA]. Data source: Tavoni et al., 2013. Data downloaded from the LIMITS Scenario database hosted at IIASA, https://tntcat.iiasa.ac.at/LIMITSPUBLICDB/dsd?Action=htmlpage&page=about.)

parking space. Workplace charging and public chargers can fill gaps where at-home charging is unavailable or insufficient for the length of a given trip. Policies to promote EVs typically aim at helping to overcome one or both of these barriers, or they provide other benefits (such as access to parking or expedited travel lanes) that increase the convenience of owning an EV.

Electrifying Buses and Trucks

Both light-duty vehicles, such as cars and SUVs, and heavy-duty vehicles, such as buses or trucks, can be electrified (Figure 8-3). However, because of different market characteristics, vehicle performance requirements, and level of technological maturity, policy considerations are different for electrification of buses and trucks than for passenger cars and SUVs.

Although the technology exists to electrify intracity buses, these buses are purchased primarily by government transit agencies, which are not responsive to the same pressures and incentives as consumers. Electric buses that draw current from guidewires were in use as early as the 1910s,[7] but they need overhead infrastructure. Battery electric buses are a more recent invention, but they have a growing market share, with more than 200,000 electric buses sold in China alone in 2015 and 2016.[8] In Shenzhen, the government just completed a full transition of its bus fleet, all 16,359 of them, to electric buses.[9]

The driving cycle of intracity buses has two key characteristics that make them highly compatible with battery electric drive trains: the prevalence of stop-and-go driving, which maximizes the value of regenerative braking, and regularly scheduled idle periods at specified locations, which offer an opportunity for high-voltage rapid recharging. U.S. electric bus maker Proterra estimates that half of new bus sales to transit agencies may be electric by 2025, and all may be electric by 2030.[10] The development of this market can be facilitated by deployment-driven cost declines and by government procurement policies, which can take into account the air quality and climate change benefits of electric buses when making new bus purchases.

Electric trucks have been in use through the 2010s and have been focused on intracity uses, such as delivery vehicles that make many stops, as well as trucks used for port and airport operations. The driving cycles of these trucks share many of the characteristics of the driving cycles of intracity buses, making them good fits for electrification. Long-haul electric semi-trucks are still in their infancy and may someday achieve long ranges through systems involving leased batteries that are exchanged at battery swapping stations at truck stops

Figure 8-3. Battery electric bus in service in Adelaide, Australia. (Photo by User: Orderinchaos (Own work) [CC BY 3.0] via Wikimedia Commons.)

along major routes.[11] Increasing the penetration of electric trucks may depend heavily on policies that support research and development efforts that will deliver cost declines more rapidly (discussed in Chapter 14).

Electrifying Motorbikes

In some countries, huge numbers of people ride motorbikes. Although motorbikes tend to be fuel efficient because of their light weight, they also tend to be heavily polluting because of their reliance on simple but dirty two-stroke engines and lack of emission control equipment, such as particulate filters.

Governments may focus on promoting electric bikes as a substitute for traditional motorbikes, using the same policies discussed in this chapter in regard to passenger cars and SUVs.

The remainder of this chapter concerns primarily the electrification of passenger cars and SUVs.

Electrifying Passenger Cars and SUVs

This section discusses the policy options for increasing passenger car and SUV electrification.

Subsidies and Rebates

One of the most effective policies a government can use to encourage EV adoption is to directly subsidize the purchase of new EVs by businesses and consumers. For example, in the United States, the federal government offers a tax credit of up to $7,500 per electric or plug-in hybrid vehicle, and some states and utilities offer their own rebates on top of this amount. For example, Colorado offers the highest state tax credit, of $5,160 per EV.[12] Subsidies are most effective when reflected in the sticker price seen by purchasers, so that they take the subsidy into account when making purchase decisions.

Subsidies are generally intended as a temporary measure, to help make the upfront purchase price of EVs more comparable to that of gasoline vehicles. Subsidies may be phased out as research on battery technology brings down the cost of batteries and increased EV production lowers per-unit manufacturing costs through economics of scale. EV batteries have followed a price reduction pathway similar to that followed by other technologies, and battery costs are projected to decline by almost half over the next 5 years, as shown in Figure 8-4. EV battery costs have dropped more quickly than most industry experts and modelers projected over the past 10 years.[13]

Cost declines do not happen automatically with time. They depend on the increasing deployment of technology delivering economies of scale and also learning by doing: producers gaining experience making EVs and finding opportunities where small changes to processes can reliably reduce inputs, increase outputs, improve quality, and so on. Subsidies help promote deployment in the early years, thus making possible the very cost declines that allow removal of subsidies later. Although subsidies may appear to be an expensive policy in the short term, they may save money for society in the long run by

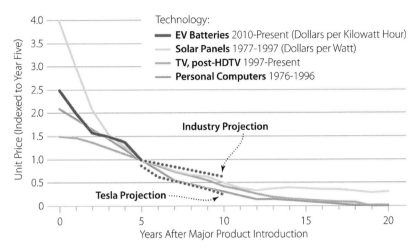

Figure 8-4. The price of electric vehicle batteries is expected to continue falling, as with other comparative technologies after product introduction. (Martin R. Cohen, "The ABCs of EVs: A Guide for Policy Makers and Consumer Advocates." Data reproduced with permission from the Citizens Utility Board.)

bringing down costs and helping EVs to more rapidly achieve cost parity with or become cheaper than gasoline vehicles.

In addition to subsidies for the purchase of EVs, government may offer subsidies in the form of reduced bridge or roadway tolls, or they may provide free electricity for charging vehicles at public charging stations. For example, the San Francisco–Oakland Bay Bridge toll is $6 during commute hours for most passenger cars, but EVs (along with carpools) are charged a reduced rate of $2.50.[14] This saves roughly $900 per year for a Monday–Friday, noncarpool driver.

Expanding Charging Station Access

EV owners need access to charging stations: connections to the electric grid that deliver electricity to vehicles (Figure 8-5). There are three types of charging stations. Level 1 chargers supply U.S.-standard 120-volt alternating current (AC) and can provide a typical EV less than 10 miles of range per hour of charge.[15] Level 2 chargers supply 240-volt AC and can transfer 10–30 miles' worth of charge per hour, depending on the type of charger (there are several varieties).[16] Level 3 or "quick" chargers can currently give a 240-mile-range EV

Figure 8-5. An electric vehicle charging while plugged into power supply. (Photograph: franz12/Shutterstock.com.)

80 percent of its capacity in about half an hour. Most U.S. buildings have 120- and 240-volt connections to the grid, but level 3 charging typically requires installation of special equipment. All EVs are able to use level 1 and level 2 chargers, but only select vehicles are equipped to use quick chargers.

EV buyers who own their own homes and have a dedicated garage or off-street parking space may choose to have a level 1 or 2 charger installed. However, this option is not typically available to renters (comprising 43.3 percent of U.S. households, for example),[17] who often do not have the ability to make upgrades to their building, or those who do not have a reserved or consistently available parking spot at which to install the charger. Similarly, owners of office buildings, shopping malls, and other commercial real estate may be reluctant to install charging equipment if they do not expect a financial return from doing so (e.g., by being able to charge higher rents to commercial tenants).

Government can make the process easier by offering rebates or tax incentives to individuals, multiunit residential buildings, and businesses that install chargers in parking areas. Subsidies may be made contingent on conditions such as public access to the chargers or participation in utility efficiency or smart charging programs.[18] Local governments may change building codes to require that new or significantly renovated buildings above a certain size include a certain number or share of chargers in garages or parking lots.

Government may also work with companies to deploy chargers in public areas, such as city-owned parking lots or alongside streets. There are several ways to pay for these stations. For example, the government may solicit bids from private companies and directly fund the charging stations.

Public utility commissions, which regulate electric utilities, can also permit utilities to recover the costs of building a specific number of charging stations in the rates they charge to customers. Utilities can expect to recoup their investment in charging infrastructure through sales of electricity to vehicles and by using vehicle charging as responsive demand, to accommodate instances of short-term oversupply of power to the grid. Three utilities in California—San Diego Gas & Electric, Southern California Edison, and PG&E—received approval from the California Public Utilities Commission to deploy thousands of charging stations (including some in multiunit dwellings and at businesses). Utility regulators and stakeholders are beginning to recognize that transportation electrification can benefit all utility customers because the additional utility company revenue from EV charging can support operation and maintenance of the existing distribution infrastructure, helping reduce the need for future electricity rate increases.[19]

Either the direct incentive or utility EV program approach allows government to impose conditions on the program, such as a requirement to deploy a certain share of chargers in disadvantaged communities or a requirement to bill electricity for charging at special rates, which are often important considerations.[20]

Roadway and Parking Privileges

Another way government may increase the attractiveness of EVs is to provide nonmonetary benefits to EV owners, such as access to special travel lanes on highways during peak commuting hours. Similarly, the government may make special parking spaces available only to EVs—often the ones with charging infrastructure. These policies increase the convenience of driving and parking an EV.

The benefits of access to special travel lanes and parking spaces derive from the limited number of vehicles that may use these services. As EVs achieve higher market share, it may be necessary to withdraw roadway and parking privileges.

Consumer Education

Government or utilities may provide consumers with information about EVs, such as projected savings on fuel and maintenance, charging options, public databases of charging station locations, reduced air pollution, and available incentives.[21] There is a need for high-quality, objective information about EVs. Dealers often steer buyers away from electric cars[22] because they need almost no maintenance,[23] and dealers make three times more profit from servicing vehicles than from new car sales.[24] Some dealers have refused to allow prospective buyers to test-drive EVs,[25] and others have played up perceived drawbacks such as range limitations[26] and withheld information on the benefits of EVs. Government may provide information through partnerships with community and environmental organizations, awards for top electric vehicle sellers, print or online media, partnerships with utilities, and so on.

Zero-Emission Vehicle Mandates

A government may mandate that automakers sell at least a certain percentage of zero-emission vehicles or EVs. This is a technology-pushing policy, helping to ensure automakers are investing in zero-emission vehicle technologies and making the resulting products available to consumers. Automakers may comply with a mandate by making larger numbers of EVs and more models available for sale, promoting EVs via marketing or consumer education, and reducing the price of EVs until consumer demand is sufficient to allow compliance with the mandate. A zero-emission vehicle mandate may be based on tradable credits, allowing manufacturers that are unable to generate sufficient credits from their own sales to buy credits from other manufacturers. Unlike a subsidy, a zero-emission vehicle mandate does not require government expenditures, although a stringent mandate will incur costs to automakers in the near term.

California and nine other U.S. states participate in a program that mandates EV sales. Each automaker is required to sell a sufficient number of EVs to meet its zero-emission vehicle credit obligations. This translates to sales of EVs rising from 2 percent of overall sales in 2018 to 8 percent by 2025. The actual number of electric cars sold may vary, though, because automakers generate fewer credits for selling a plug-in hybrid with a gasoline engine, a vehicle that only partially runs on electricity, and more credits for a battery-EV powered exclusively by electricity.[27] (Unfortunately, partial credits were cheap and easy to attain, leading to an oversupply of credits and reducing pressure on manufacturers to deploy zero-emission vehicles.[28] The problems with oversupply illustrate the importance of careful policy design to ensure policy effectiveness.)

Zero-emission vehicle mandates are spreading worldwide. China, the world's largest auto market, has announced its own zero-emission vehicle mandate, which will require 8 percent of automakers' sales to consist of zero-emission vehicles by 2019.[29] The European Commission is considering a similar mandate program for Europe.[30]

When to Apply These Policies

Although EVs are a new technology, they hold promise for both developed and developing countries, so all countries should be using policy to prepare for and facilitate a transition to EVs. That EVs fit well in developed countries may be unsurprising: Developed countries have the money to afford new technology, and EVs will be a crucial component of efforts to decarbonize the transportation sector and achieve aggressive greenhouse gas emission reduction targets. But EVs are a surprisingly good fit for developing countries as well:

- Most developing countries do not produce large quantities of oil or petroleum fuels domestically, so they must import the fuel for gasoline- and diesel-powered vehicles, sending precious cash abroad. EVs run on electricity, most of which is produced domestically, and they go much farther per dollar spent on fuel. This provides mobility services more cheaply, and more of the money spent on energy remains in-country.
- As they industrialize, many developing countries are suffering from terrible air quality. For example, air pollution in China and India kills millions of people per year.[31] On top of the direct humanitarian and economic costs, air pollution makes cities undesirable places to live and

work. This can be a drag on the development process. EVs reduce harmful air pollution in cities. Ideally, electricity will be supplied by zero-emission technologies such as solar, wind, hydro, and nuclear, but even natural gas emits very few particulates (the most damaging type of air pollution to human health). Even if the electricity for EVs is supplied by coal, the coal plants can be located far from population centers, whereas gasoline and diesel vehicles emit pollution in the hearts of dense cities, where more people are exposed to the pollution.

- Whether or not EVs are deployed, developing countries commonly seek to build out a robust electric grid, to supply their citizens with power. EVs can use much of the same grid infrastructure and can even help contribute to grid stability; For example, by varying their charging rates based on the needs of the grids during different parts of the day, EVs can provide a valuable grid service. A developing country whose population still largely lacks access to private vehicles may be able to leapfrog internal combustion engine vehicles in favor of EVs. This might allow countries to avoid investing in costly infrastructure for gasoline vehicles, such as pipelines, refineries, tanker trucks, and fueling stations.

- EVs are more reliable and have fewer moving parts than internal combustion engine vehicles, so they may need fewer repairs over their lifetime, lowering the total cost of ownership.

The biggest concern about EVs in developing countries is cost. Although EV prices are dropping rapidly, they remain more expensive to purchase than similar petroleum vehicles. The price gap is shrinking, however, and developing countries should plan their infrastructure and their policy for the long term, when EVs will be cheaper than internal combustion engine vehicles. Nonetheless, because of differences in grid readiness and in citizen and government resources, different policies are best suited to facilitate EV deployment in developed and in developing countries.

In developed countries, policy should focus on starting broad displacement of privately owned internal combustion engine vehicles by EVs. Expanding access to charging stations and facilitating deployment of at-home chargers (e.g., by ensuring a fast, efficient, and inexpensive permitting process for home installation) is a priority. Developed countries can afford to subsidize EVs, and EV subsidies have been used in many European countries, the United States, Canada, and elsewhere. Access to high-occupancy travel lanes on highways has proven to be a major factor in consumers' purchase decisions, and EV sales mandates may be achievable by vehicle manufacturers.

In developing countries, it may be advantageous to begin by facilitating deployment of EVs in ecosystems and roles for which they are particularly well suited. For example, a government can choose to procure EVs for its vehicle fleets, and it may make subsidies or other incentives available for the electrification of corporate vehicle fleets, intracity delivery trucks, and trucks used in port operations. These are environments where purchase decisions are made by professionals who can take account of lifetime total cost of ownership of the vehicle, where the vehicles often return to a central location where they can be charged, and where vehicles have a stop-and-go duty cycle that maximizes the fuel savings from regenerative braking. Starting with EV deployment in specific ecosystems or roles also gives utilities time to prepare grid infrastructure for more widespread EV deployment.

When a developing country is ready to begin rolling out EVs to private citizens on a widespread basis, a feebate (discussed in Chapter 7) is one of the first policies to consider. A feebate can have a powerful effect on consumer choices, because it combines the effects of a tax and a subsidy, and it modifies the upfront purchase price, the most salient factor to most consumers (as opposed to the lifetime cost of ownership, because most consumers heavily discount or disregard future fuel savings). A feebate also is affordable for a low-resource government, because the feebate can be revenue neutral, unlike a pure subsidy.

Detailed Policy Recommendations

Policy Design Principles

The following policy design principles apply to EV policies.

Create a Long-Term Goal and Provide Business Certainty

When using subsidies to promote EV adoption, the subsidy rates and eligibility of different vehicles should be established and publicly announced at least several years in advance. When subsidies phase out, they should do so according to a schedule, also known years in advance. These steps prevent abrupt or unexpected changes in subsidy rates that can lead to sudden crashes in sales and harmful disruption to the automobile manufacturing industry. For example, when the U.S. state of Georgia abruptly repealed its $5,000 EV tax credit and replaced it with an annual $200 fee on EVs, registrations of new EVs plunged by 90 percent.[32] Georgia's tax credit is discussed in greater detail as the first case study in the next section.

If flexibility is necessary because of uncertainty in the rate of future technological advancement, the subsidy phase-out may be based on the cost differential between internal combustion engine vehicles and EVs of the same type (e.g., midsize cars, SUVs). In this case, the formula to be used to calculate the subsidy, rather than the ultimate subsidy value, would be publicized in advance.

Use a Price-Finding Mechanism

If a policymaker's goal is to achieve a particular level of EV sales, it may be prudent to use a price-finding mechanism to determine the lowest subsidy rate that would achieve the desired sales level. This is not how EV subsidies are usually structured today, either because policymakers do not have a specific EV deployment target or because they have a particular quantity of funding that they are willing to spend on the subsidy program, which they are not willing to exceed, even if this will cause the target to be missed.

To make the funding for a subsidy program go as far as possible, the subsidy may be targeted at buyers who wouldn't have purchased the vehicle otherwise. One approach is to target middle-income consumers rather than upper-income consumers or to limit incentives to lower-priced electric vehicles.

A price-finding mechanism may be applied to the deployment of EV charging infrastructure by soliciting bids from various providers. The company that presents the lowest-cost bid that provides the most chargers in the most useful locations may get the contract.

Eliminate Unnecessary Soft Costs

Policymakers should streamline the process of registering EVs and obtaining permits to install charging infrastructure in homes and businesses. To install a car charger, some cities require a new owner to submit a building permit, floor plans, electrical service load calculations, and other documentation, along with associated application fees.[33] Utilities should ensure the electric grid is able to handle the load from EVs and precertify neighborhoods or other large areas, reducing or eliminating the need to perform load calculations on a case-by-case basis. Government may provide basic templates explaining key choices, such as where to locate a charger relative to the parked vehicle and whether to install a second electric meter (in places where a special, lower electricity rate is charged for electricity used to charge EVs).[34]

Apply the Policy to the Smallest Set of Actors That Achieves 100 Percent Coverage of the Market

If subsidies are used, they should be available to all EVs that meet certain performance requirements (such as a minimum battery size). In the United States, the federal EV tax credit phases out on a per-manufacturer basis once a manufacturer has sold 200,000 qualifying vehicles. Although no manufacturer has reached this limit yet, several manufacturers will soon,[35] but the tax credit will remain available to other manufacturers for many years. This policy design may encourage each automaker to consider entering the EV market. However, it will lead to fewer subsidies going to manufacturers who are leading the market, which are generally the early movers that offer the least expensive or best-quality products. It is better for the market and for emission reduction if policies are manufacturer neutral and reward the best technology.

If 100 percent of the market cannot be targeted (perhaps because a country is in too early a stage of development), then high-impact use cases should be targeted first, such as short-haul commercial trucking, corporate or government vehicle fleets, taxis, or private vehicles in the most developed urban areas with better grid infrastructure support. This will help to lay a foundation to facilitate increasing penetration of EVs over time.

Other Design Considerations

Distributional Effects

Policies designed to maximize the adoption of EVs may have unintended side effects, such as exacerbating income inequality or favoring urban over rural residents. In California, households in the top income quintile received 90 percent of all EV tax credits,[36] a transfer of general tax revenues to the highest earners. Policies that promote EVs also tend to have less impact on rural areas, where travel distances are longer and charging infrastructure is scarcer.

One way to limit the impact of EV promotion policies on inequality is to allow subsidies only for buyers (or lessees) with income below a certain cap or to limit subsidies to vehicles priced cheaply enough that they are available to the majority of citizens. Although EVs may be unlikely to become popular in rural areas in the near term (because of range and cost concerns), governments can prepare for eventual rural EV adoption by extending charging networks into rural areas. The best spots may be along major travel corridors such as highways, so that the chargers can be used by either local residents or passing motorists, improving their economics.

Additionally, it is worth noting that in some cases, policies that benefit primarily higher-income earners in the short term can benefit all consumers in the longer term. Early adopters, such as the first EV drivers in California, assumed the risk of buying an untested technology for which the supporting technological infrastructure—such as chargers—had not been widely deployed. This provided a market for EVs, helping manufacturers innovate and pushing EV prices down their learning curves, yielding cheaper products for the broader market in the future. Early adopters also contribute to social acceptance of EVs as capable and reliable vehicles, laying the groundwork for EV market development.

Case Studies

Georgia EV Tax Credit

In the 1990s, Atlanta was out of compliance with federal air quality standards for ozone, and vehicle emissions were primarily responsible.[37] In 1998, the legislature passed a $1,500 tax credit for alternate fuel vehicles, which was increased to $2,500 for all low-emission vehicles and $5,000 for zero-emission vehicles over the next 3 years. The tax credit applied to buyers and first lessees of EVs. At the time, the bills were uncontroversial.[38]

As years passed, EVs became more widely available and declined in cost. The tax credit was successful at helping EVs gain a foothold in Georgia, particularly in Atlanta, where the range limitations of early EVs were less problematic than in rural areas. By mid-2015, Georgia had more EVs in service than any other U.S. state except California.[39]

EV costs declined and availability increased during this period. In 2015, one of the least expensive EVs, the Nissan Leaf, had a sales price of $30,000, or $22,500 after the $7,500 federal tax credit. Dealers began offering 2-year leases for as little as $199 per month. Georgia's $5,000 tax credit, spread over 24 months, could cover the entire cost of the lease.[40]

The ability to essentially own an EV for free drew the ire of legislators representing rural portions of the state, who portrayed the policy as "giving free cars to Atlanta yuppies."[41] EV advocates recommended cutting the tax credit in half, then phasing the remaining half out over 3 years.[42] Going even further, the state passed a bill that terminated the tax credit abruptly and also imposed an annual $200 fee on EVs, the steepest such fee in the country.[43] As a result, the market crashed, with registrations declining by 90 percent.[44]

This was a desired outcome for Rep. Chuck Martin, the legislator who sponsored the bill to repeal the tax credit, who stated that the drop-off in sales "vindicates that the credit needed to be removed."[45] However, it is not a good result for the future of EVs in Georgia, nor for mitigating climate change.

Several lessons can be drawn from Georgia's experience. First, subsidies must be revisited at known intervals to keep up with technological change. Georgia's subsidy rate, set to $5,000 in 2001, became overly generous more than a decade after it was enacted.

Second, the policy failed to account for distributional effects (described earlier). For example, if some of the benefits had accrued to rural areas, it might not have garnered so much opposition.

Third, abruptly ending a subsidy policy can cause a dramatic shock to the EV industry. Financial incentives should be phased out according to a multiyear schedule (potentially linked to the cost differential between gasoline vehicles and EVs, as noted earlier).

Fourth, when a subsidy is offered to a lessee rather than a buyer, it may be prudent to spread out the value of the subsidy over time. When the entire subsidy is provided to the first lessee, this makes the car cheap to lease for the first lease term, but the lessor may have difficulty leasing or selling the vehicle when that term expires because the subsequent lessee or buyer will not get any tax credit. A policy that (for example) awards only 20 percent of the tax credit to a lessee per year would ensure that the benefits are spread out among the first 5 years' worth of lessees (or buyers, if the car were sold within the first 5 years).

China's EV Subsidies

In 2009, China adopted an ambitious plan to become the world leader in EV technology and manufacturing. The plan included large subsidies for EVs sold in the domestic market, and the central government ordered State Grid, the Chinese electricity system operator, to install charging stations in major cities.[46]

Chinese manufacturers began producing EVs in large numbers. To keep EV prices as low as possible, an important goal in the price-sensitive Chinese market, companies designed EVs with poorer performance characteristics than those sold in many other countries. A typical Chinese EV may have a range of

Figure 8-6. The Roewe E50 is an electric vehicle released by Chinese manufacturer SAIC Motor in 2013. (Photo by Navigator84 [Own work], CC BY-SA 4.0, via Wikimedia Commons.)

120 miles and a top speed of 60 miles per hour (Figure 8-6).[47] However, these limitations have been acceptable to Chinese consumers, who drive primarily short distances in cities, where top speeds are limited by traffic.[48]

A top selling model in China in 2017, the Chery eQ, costs 60,000 yuan ($8,655) after subsidies. Without subsidies, the price would be 160,000 yuan ($23,080).[49] The Chinese government also exempts EV owners from vehicle ownership taxes,[50] which can be very expensive, sometimes costing more than the vehicle itself.[51]

Because of the availability of these extremely cheap EVs, the market boomed. Today, China possesses 38 percent of the world's EVs, more than three times as many as the United States.[52] 2016 sales were 507,000 units, more than twice the sales in Europe and more than four times the sales in the United States.[53]

The subsidies have "helped cultivate a gold-rush mentality," with more than 200 manufacturers, some of whom have little expertise in building high-quality vehicles, hurrying to sell hastily constructed EVs in order to pocket the subsidies.[54] The value of the subsidies also enticed some manufacturers to commit fraud by illegally registering vehicles, using smaller batteries in production than in testing, and falsifying sales figures.[55] To remedy these problems and force manufacturers to improve the quality of the vehicles they sell, the Chinese government has begun improving enforcement, mandating better energy consumption and driving distance in order for vehicles to qualify for subsidies,[56] and will phase out its EV subsidies by 2020.[57] The central government has also capped the level of subsidy that provincial governments may offer for EVs.[58] Going forward, China is considering instead using a zero-emission vehicle mandate modeled on the California mandate[59] discussed earlier, which will obligate vehicle makers to continue the transition to EVs without providing financial reward to firms making low-quality products.

Improving EV quality will be crucial for Chinese manufacturers to expand their sales into foreign markets, a goal of GAC Motor and BYD, two of China's biggest EV manufacturers.[60]

Using a combination of generous subsidies and government-mandated charging infrastructure buildout, China was able to become the world leader in EV deployment in half a decade. By taking steps to improve vehicle quality and withdraw subsidies gradually, according to a schedule announced years in advance, China is well set to avoid a dramatic crash in EV sales, such as that experienced after the sudden withdrawal of subsidies in Georgia. If the Chinese policies had required better performance in order to qualify for subsidies earlier in the lifetime of the subsidy program, they might have avoided the proliferation of companies making low-quality EVs, saved government money, and prepared Chinese EVs for export to developed markets more quickly. Despite this flaw, the Chinese program has been successful at achieving its goal of making China a world leader in EV manufacturing.

Conclusion

Electrification of on-road vehicles will be an important part of decarbonizing the transportation sector. Although EV technology has come a long way, it still needs government policies to accelerate adoption, at least in the near and medium term. Key techniques for policymakers include rebates and subsidies, development of charging infrastructure, and consumer education. It is crucial that these policies be designed with a long time horizon, that subsidy rates keep up with technological progress, and that they be phased out gradually according to a schedule or formula set years in advance. Programs in Georgia and China show the potential for policy to achieve success and specific pitfalls to avoid. With well-designed regulatory incentives and cost declines from advancing technology, quiet, zero-emission EVs can quickly become a common sight in cities worldwide.

CHAPTER NINE

Urban Mobility Policies

Many cities are clogged with traffic, a problem that will only get worse as cities grow and the world continues to urbanize. In 2010, 76 percent of people in developed countries and 46 percent of people in developing countries lived in urban areas; by 2050, these percentages are projected to grow to 86 percent and 64 percent, respectively.[1] Traffic congestion in cities causes severe impacts on residents and on society at large, including increased greenhouse gas and conventional pollutant emissions, lost time and productivity, and increased transportation costs. Infrastructure is long lived, so poor choices made today will set energy use patterns and affect residents for generations to come.

Smart policies to enable alternative forms of urban mobility and reduce the number of vehicles on the roads can improve quality of life while dramatically cutting transportation sector emissions. Several types of policy can contribute to this goal. Not every policy will be well suited to every city (e.g., some cities are rapidly growing, whereas others are already largely built out), so each city should enact the policies that best suit its physical, economic, and political circumstances. This chapter discusses key policies in a global context. For country-specific policies, see "12 Green Guidelines,"[2] a paper authored by Energy Innovation and China Development Bank Capital, or, in Chinese, the much more detailed handbook *Emerald Cities*.[3]

Urban mobility policies can contribute at least 2 percent of cumulative emission reductions, in line with a two-degree scenario by 2050 (Figure 9-1). It's worth noting that these policies can be difficult to measure, and reductions could be significantly higher.

Policy Description and Goal

Well-designed cities are people oriented and provide robust transportation options, whereas poorly designed cities can create gridlock and poor air quality. The key policies for excellent urban mobility are well-designed, well-funded

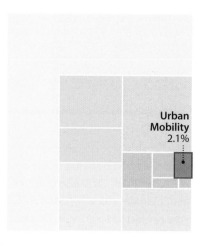

Figure 9-1. Potential emission reductions from urban mobility. (Analysis done using data with permission from the International Institute for Applied Systems Analysis [IIASA]. Data source: Tavoni et al., 2013. Data downloaded from the LIMITS Scenario database hosted at IIASA, https://tntcat.iiasa.ac.at/LIMITSPUBLICDB/dsd?Action=htmlpage&page=about.)

public transit; mixed-use and transit-oriented development; measures to facilitate biking and walking; compact and infill development; and vehicle control. The value and design of these policies are likely to vary with a city's size. Key considerations for urban mobility policies include regional coordination, acquiring funding for major infrastructure projects, and overcoming political opposition. Case studies include bus rapid transit in Guangzhou, China; transit-oriented development in Curitiba, Brazil; support for bicycles in Copenhagen, Denmark; and congestion pricing in London, UK.

Well-Designed, Well-Funded Public Transit

For cities of all sizes, public transit is the backbone of the transportation system, particularly during busy commute hours when roadways are at or over capacity. For public transit to succeed, it must be a first-class option—a travel mode people prefer because it provides a better experience and a higher level of service than driving alone in a private vehicle.

Metro (subway) systems can provide a high level of service, but they are very expensive to construct, which can be prohibitive for many cities. Bus rapid transit systems (Figure 9-2) can provide similar benefits and performance to metro lines and cost dramatically less—such systems are more than 90 percent cheaper than an underground subway.[4]

Good bus rapid transit systems are not just a collection of "express" buses with limited stops. A high-quality bus rapid transit system will have station platforms with fare gates so riders may pay when entering the station, not as they board the bus, thereby speeding up the boarding process. Stations should be built to bus boarding height. The buses should have whole walls of doors, like a subway car. With passengers already in the station and the bus open for

Figure 9-2. Bogotá's bus rapid transit system features dedicated bus lanes and preboarding fare collection. (Photo by Karl Fjellstrom, Far East Mobility, CC BY-SA 4.0, via Wikimedia Commons.)

boarding with no stairs and many doors, boarding and exiting are just as fast as a subway.

Bus rapid transit buses should have their own dedicated lanes so they are not slowed by traffic, and they should carry transponders that optimize traffic signals to accommodate approaching buses. Both bus rapid transit and traditional public buses should use real-time tracking to provide passengers with information on arrival and departure times. This is especially important in small and midsize cities, where public transportation options tend to run less frequently given the fewer number of passengers.

More broadly, transit systems should be coordinated so passengers can easily switch lines and modes. Multimodal systems are becoming even more important as bike-share, car-share, autonomous vehicles, bus rapid transit, light rail, metro, and other modes are all becoming part of the transportation mix. Integrated fare systems, shared structures (e.g., a single terminal for different modes), and parking management are important elements of a well-integrated transit system. Transit centers should also contain bike-sharing stations and bicycle parking and be designed for walkability to facilitate switching to non-motorized modes of transport.

Coordinating different transit modes is essential for getting the whole system to work, especially including connectivity to public transit as the backbone of the system. Information technologies can also improve the transit experience. Real-time data on vehicle locations can reduce the time passengers wait for a transit vehicle and can assist in optimizing dispatch. A universal smart card can streamline passengers' payment across all regional transit systems and vehicle types.

Increasingly, ride-share applications such as Uber and Lyft are becoming a central part of transit in cities. Although it is important for cities to plan and optimize around these growing forms of transportation, they are not and should not be thought of as replacements for traditional transit systems. Relying heavily on these services in lieu of other transit options, such as buses or light rail, can lead to large negative impacts. For example, in New York City, ride-sharing apps have made traffic much worse.[5]

Mixed-Use and Transit-Oriented Development

The goal of urban mobility is not to allow people to cover the maximum number of miles, whether by bike or car, but to create access to what people need every day. Smart urban planning can reduce the number of trips people make and, when people do travel by vehicle, can shift some of those trips onto transit vehicles. Mixed-use zoning locates residential, commercial, and other uses (e.g., community services, healthcare, education) in the same area. Residents of a mixed-use neighborhood are able to reach many different types of amenities without a vehicle. This reduces the number of trips needed for tasks such as grocery shopping, dining at a restaurant, visiting a bank, and so forth. Some employees of the businesses in mixed-use areas will choose to live in the area, eliminating commute trips as well. Mixed-use zoning also creates a vibrant, walkable streetscape and can be particularly beneficial to elderly and disabled people who may have trouble traveling long distances.

Transit-oriented development involves zoning for high densities near metro stations and along transit corridors. Many people are not willing or able to walk far in order to get to trains; the walking distance to the nearest high-capacity transit center should be no more than 1 kilometer (a little over a half-mile).[6] By ensuring most residences and places of employment are located near convenient transit options, a city can minimize the number of trips people choose to take using private vehicles. Density should be matched to transit capacity, with the highest densities near large stations that can accommodate the resulting flux of travelers. Areas around stations should be walkable and include mixed uses, because high foot traffic around transit stations can benefit the nearby retail outlets.

Development should be inclusive. Specifically, this means ensuring connections to affordable housing, which is often placed farther out of the city center with less access to transit for the people in greatest need of service.

Biking and Walking

The most vibrant and livable cities around the world are those that boast an attractive environment for walking and biking. These nonmotorized modes of travel emit no pollution and offer public health benefits by increasing people's physical activity. Their space efficiency also reduces the amount of land needed for parking lots and structures, parking alongside roadways, and travel lanes. This allows more of that space to instead be devoted to people (parks, shops, offices, and so on).

Cities should include dedicated biking and walking paths that are protected from motor vehicle traffic. Certain streets, particularly those lined with shops and restaurants, can be changed to pedestrian boulevards that limit or prohibit the use of cars. Sidewalks should be wide and feature amenities such as trees, benches, and high-quality lighting to increase the appeal of walking. In some cases, city transportation infrastructure or right-of-way that is no longer in use can be converted for biking or walking, such as Chicago's 606 elevated trail (Figure 9-3) or New York City's High Line.

Street layout also has a strong effect on walkability (Figure 9-4). Interconnected street networks with small block size and smaller streets maximize a neighborhood's walkability and bikability. Layouts with wide streets and large block sizes (as in many Chinese cities), as well as poorly interconnected street layouts with winding roads and cul-de-sacs (as in many American suburbs), increase the average distance to reach a destination. They also tend to funnel more cars onto fewer streets, which increases traffic congestion despite the greater width of these arterial streets.

Compact and Infill Development

A city sprawls outward when developers build on the city's periphery, often converting greenfields (e.g., farm fields, pasture, and wilderness) into roads, houses, and buildings. Because the cost of land on city outskirts is low, developers tend to use larger lot sizes, building widely dispersed houses with large yards. These areas can lead to car dependence, because many homes are too far away to easily walk or bike to common destinations.

Developers should instead be encouraged to build on unused or low-value

Figure 9-3. Chicago's 606 multiuse trail was previously an abandoned rail line. (Photo by Victor Grigas [Own work], CC BY-SA 4.0, via Wikimedia Commons.)

land already within the developed footprint of the city, such as building on empty lots or surface parking lots, replacing abandoned or low-value structures with taller and better ones, and renovating historic buildings that have fallen into disuse.

One of the strongest tools to achieve compact and infill development is an urban growth boundary: a line encircling a city beyond which development is not permitted (except for certain uses, such as farms and parkland, which are not part of the city's footprint), like the Green Belt in London, UK.[7] The urban growth boundary can be expanded periodically by an act of the city or regional

A. Large block size   **B. Poorly interconnected streets (cul-de-sacs)**   **C. Small block size with well-connected streets**

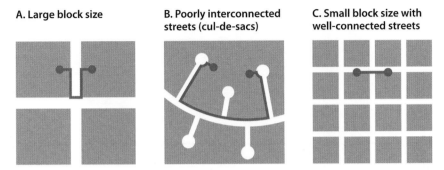

Figure 9-4. The effect of street layout on average travel distance.

government if opportunities for infill development grow scarce. Other tools to promote infill development include reducing impact fees (fees levied by cities on developers of new projects to fund expansion of city services, such as extra police, school, and firefighting capacity) and expediting the application process and other paperwork for infill projects.

Vehicle Control

Vehicles cause large social impacts, including harms to public health, traffic accidents, and lost time and productivity due to traffic congestion. Vehicle control policies seek to reduce the number of vehicles on the road by either discouraging car owners from driving in dense urban areas (by making it more inconvenient or expensive to do so) or optimizing vehicle routes to reduce congestion.

The simplest and cheapest form of vehicle control is parking management. Cities may reduce the quantity of parking or charge higher parking fees. This allows more land to be devoted to other uses and imposes a financial penalty for using a car to reach a busy part of the city. With new technology, parking fees can vary based on usage, so it becomes more expensive to park as fewer parking spaces remain available. This helps ensure parking capacity is rarely, if ever, exhausted. Parking management is easier and cheaper to implement with various sensors and information technology systems than many other forms of car control, and it can create revenue for other transport investments.

Congestion pricing imposes a fee on drivers entering the city center or other crowded areas, particularly at peak hours. Crowded roadways may also be

Figure 9-5. Cornelius, Oregon's urban growth boundary has encouraged infill development and protected surrounding open space. ("Cornelius, OR," Map. Google Earth. Google, September 12, 2016.)

managed with congestion pricing—effectively converting them into toll roads whose fees vary with congestion level.

Another form of vehicle control involves limiting the number of vehicle registrations. Permits can be distributed by lottery, which is fairer for all socio-economic classes, or by auction, which generates revenue the city can spend to improve urban mobility for lower-income groups and other residents who do not use cars.

Cities may also impose zoning restrictions capping the number of parking spaces in new developments at one (or fewer) per household so these developments will attract residents who prefer to live with just one car (or none at all).

When to Apply These Policies

Every urban area should be integrating forward-thinking urban mobility goals into its development plans. Designing policy correctly from the outset can prevent urban sprawl and lead to a pattern of infrastructure development and urban form that will promote compact growth, maximize walkability, and increase quality of life for residents (Figure 9-5).

The right policies to promote smart urban development depend on the physical characteristics, size, and growth patterns of the city or region where the policies are to be enacted. The policies recommended in this section use population figures that are intended as rough guidelines. In practice, different population levels, density, transit demand, and income level may make some policies more cost-effective than others. The following policies also apply best to towns or cities that are growing or developing in population or economic size, rather than static or shrinking towns and cities.

Towns and Cities of All Sizes

All towns and cities should develop and maintain comprehensive land use and transportation plans. Plans help ensure that development occurs in a thought-out way, and they can concentrate development along existing or future transit corridors. The process of developing a plan should involve public input, and it can increase the local government's confidence that its direction and vision are supported by most residents. The plan can then serve as a guide for zoning decisions, for public infrastructure construction, and for private residential and commercial development.

Towns and Small Cities (40,000–200,000 People)

A town should establish an urban growth boundary early in its development. Open lands around the town may still be plentiful, allowing a boundary to be drawn with less political contention and less chance of cutting off "islands" of preexisting development beyond the new boundary. Encoding infill and high-density development into the town's DNA from the beginning will help set residents' expectations about the character of the town and facilitate the establishment of a vibrant and walkable downtown district.

At this size, a town is too small to support the investment necessary for urban light rail infrastructure, unless it is situated near a larger city and can be served by an extension of that city's rail system. Towns at the smaller end of the scale may be too small to justify a bus rapid transit system, because it may be possible to quickly traverse the whole town with regular bus service. A town should start with regular urban bus service. As it grows, one or two bus rapid transit lines may be established, which serve the main corridors of the town, including the central business district and the densest concentration of residential neighborhoods.

The most cost-effective period in a town's lifecycle for the government to secure rights-of-way for future infrastructure projects is when the city is still small and land values are still low. Forward-thinking acquisition of rights-of-way can allow the future development of surface rail infrastructure, which is much cheaper per mile than underground or elevated trains. This is also the best time to build biking and walking paths (Figure 9-6), which are inexpensive and can be especially useful for smaller towns, which can often be traversed effectively by bicycle.

People who live in towns of this size often need to drive elsewhere for services, so car control measures, such as limiting the number of vehicle permits or residential parking spaces, are not appropriate. Instead, local government should zone commercial areas, especially the central business district, to encourage pedestrian-friendly storefronts positioned against the sidewalk, while disallowing or at least discouraging strip malls with large surface parking lots, big-box retail, and drive-through restaurants. Parking in the central business district should be provided by city-owned parking structures, which charge a modest fee for parking, rather than by situating parking spaces in front of every shop and restaurant.

This is a crucial time in a town or city's lifecycle to begin encouraging compact and mixed-use development. Cities should zone for several-story buildings downtown, featuring ground-floor commercial space below multifamily residences. They should include aesthetic streetscape improvements such as trees, benches, and decorative lighting and should integrate parks and greenspace into the town, including its central business district.

Midsized Cities (200,000–500,000 People)

At this size, a city should make a serious and sustained commitment to high-quality, dedicated-lane bus rapid transit service or a light rail system. Devoting lanes to bus rapid transit or making space for rail lines and stations becomes more expensive and difficult at larger sizes, especially if a sprawling development pattern has been established (e.g., in Atlanta, Georgia and Nashville, Tennessee). The benefits of public transit should be clearly communicated to residents, including economic returns, public health, and less time spent waiting in traffic.

This is also an appropriate size for a city to commit to transit-oriented development projects featuring high-density, mixed-use buildings in proximity to light rail stations and other transit hubs (Figure 9-7). If an urban growth boundary is in use, it should be expanded only gradually and deliberately. Infill development can be encouraged through floor area ratio bonuses and expedited permitting.[8] Building upward, not outward, should be established as part of the city's character to reduce future opposition to development at the density

Figure 9-6. The central business district in Breckenridge, Colorado. (Photo courtesy of Pixabay via CC0 Creative Commons, Breckenridge, Colorado, Town, City, Photograph, accessed January 10, 2018.)

needed to maintain a city's walkability, livability, and affordability as it grows. Building height limits are usually a bad idea.

A city should continue to develop biking and walking paths and, at this time, also consider establishing pedestrian-only streets, which can create vibrant areas for residents to visit and shop.

Large Cities (500,000–2 Million People)

At this size, a city needs a robust, well-developed public transit system. Investments in light rail and bus rapid transit systems may now be augmented with traditional metro (heavy rail) services. In dense areas, this may include elevated or underground tracks. Tunnel boring technology has become cheaper in recent decades and allows deeper tunnels and stations that do not follow the street grid, minimizing disruption during construction.[9] Planning for continued growth should be heavily influenced by available and anticipated public

Figure 9-7. Transit-oriented development in Plano, Texas. (Photo by David Wilson from Oak Park, Illinois, CC BY 2.0, via Wikimedia Commons.)

transit routes, which will increasingly serve as the city's backbone. As the number of public transit and nonmotorized travel options grows, it is increasingly important for new projects to be designed and built to connect with existing forms of transportation (Figure 9-8).

Aside from parking fees, which can work in smaller towns, this is the city size at which car control measures start to become viable. Cities should start

Figure 9-8. The Volkstheater Metro Station in Vienna, Austria. (Photo by Leandro Neumann Ciuffo [Estação de metrô] CC BY 2.0, via Wikimedia Commons.)

limiting the amount of parking provided with new residential developments, particularly in denser areas that are well served by transit.

Cities should continue to encourage infill, mixed-use, and transit-oriented development through appropriate zoning, expedited permitting, and floor area ratio bonuses for buildings in high-priority areas.

Megacities (More Than 2 Million People)

At this size, some of the strongest car control policies become viable, including limits on the overall number of vehicles that may be registered, as well as taxing all vehicles entering the city center or central business district during peak hours. Bus rapid transit and light rail systems continue to be helpful, but they are likely to become insufficient, so it will be necessary to provide a high-capacity metro system capable of moving hundreds of millions of riders per year.

Tax-increment financing may be particularly useful to fund transit expansion. Tax-increment financing is a funding mechanism in which future

Figure 9-9. Café seating along the Chicago Riverwalk in Chicago, Illinois. (Photo by Serge Melki from Indianapolis, USA [Café by the river—Chicago], CC BY 2.0, via Wikimedia Commons.)

increases in property tax revenues in a given district are devoted to paying off debt incurred for a civic project in that district. Because projects funded with tax-increment financing can increase the property values of nearby buildings, and megacities are dense and have high property values, even a small percentage increase in property values may generate a substantial amount of revenue.

Acquiring land for biking and walking paths is often very expensive in megacities, but cities can impose requirements on private developers to provide paths or greenspace as part of their developments. For example, Chicago has expanded its pedestrian riverwalk in part by requiring all buildings constructed along the relevant section of the Chicago River to extend the riverwalk across their property (Figure 9-9).

Detailed Design Recommendations

Policy Design Principles
The following policy design principles apply to urban mobility policies.

Create a Long-Term Goal and Provide Business Certainty
Cities should not leave urban form to chance: They should draft forward-looking urban plans that will result in smart development and meet the needs of residents. Putting guidelines in place early will pay dividends by sustaining (and even accelerating) economic development and by guiding that development in a pattern that is consonant with sustainability and the city's values. Plans should account for long-term projections of population growth to ensure that investments made today will serve the city's future needs and will not lock in unfavorable energy use patterns. Developers should have clarity about what they can expect from the city, including any rewards for building in accordance with the plan. For example, such a reward might be an increase in maximum floor area ratio for transit-oriented developments.

Price In the Full Value of Negative Externalities
Vehicle traffic causes substantial social harms. When making investments in transit, enforcing congestion charges, and levying vehicle registration and parking fees, the city should remember that these interventions help offset some of the costs drivers cause that are now borne by society. Pricing should reflect social harms at multiple scales: global (climate change), regional (public health damages), and local (traffic congestion).

Eliminate Unnecessary Soft Costs
Particularly for projects that are in line with the city's plans, such as transit-oriented, high-density, mixed-use developments, the city should expedite permitting, minimizing uncertainty and cost for developers. Public transit

systems should be coordinated and constructed with neighboring towns and cities, resulting in a unified regional transit system instead of a patchwork of local systems.

Zoning ordinances should be used to highlight areas for high-density, mixed-use development. Good zoning can help developers avoid the costly and time-consuming process of obtaining zoning variances for particular projects.

Additional Design Considerations

Because of the diversity of urban mobility policies, cities must take many considerations into account, but the thorniest ones tend to fall into three categories: the need for regional coordination, the need to obtain funding for infrastructure projects, and the need to prevent or overcome political opposition to the urban mobility policies. Each will be discussed in turn.

Regional Coordination

Most towns and cities are near other towns and cities. The effectiveness of some of the urban mobility policies discussed in this chapter can be compromised if they are adopted by only one town and ignored by others. For example, an urban growth boundary is not helpful if the resulting greenspace is gobbled up by a neighboring town and used for low-density development. Similarly, transit systems can best serve the largest number of people if the needs of residents throughout the region are taken into account in planning lines and routes.

The best way to ensure regional coordination is through a higher-level authority, which can mandate that towns and cities work together to achieve specific goals. For example, Oregon requires each of the state's cities and metropolitan areas to establish an urban growth boundary. The Portland area has a single Metro Council that reviews and adjusts the urban growth boundary every 6 years.[10] California's Sustainable Communities and Climate Protection Act of 2008 requires each of the state's metropolitan planning organizations to prepare a "sustainable communities strategy." These plans cover land use, housing, and transportation, and they must indicate how each region will achieve specific greenhouse gas reduction targets.[11]

In areas where a higher-level authority does not exist and cannot be easily created, political leaders and urban planners from towns and cities throughout a region may still come together in a looser coalition to establish shared goals and create regional plans.

Funding

Some of the urban mobility policies—particularly those involving the construction of public transit, walking and biking paths, and streetscape improvements—are typically funded by the government, often by issuing bonds. Public transit projects easily pay for themselves in social benefits: One transit study in Chicago showed a 21 percent annual rate of return, which rises to 61 percent if coupled with transit-oriented development.[12] But whatever rate of return cities might expect, they often find it difficult to come up with the necessary initial investment. It goes without saying that once a project is funded, it is critical that the project continue to receive funding in each budget cycle. Failure to continue adequately funding transit projects can result in significant degradation of these projects over time. Governments may consider a variety of approaches to obtain the necessary funding:

- Prioritize funding for public transit projects over those that benefit private vehicles. Public transit improvements, including for nonvehicle transit (e.g., bike lanes), reduce the amount of infrastructure needed to support vehicles. Funding that would have gone to building and maintaining new roads may instead be redirected to transit projects.
- Issue bonds for a specific project. In some regions, the government will issue bonds to raise revenue for infrastructure projects. Bonds are typically paid back through taxes (e.g., sales tax or property tax) or through revenues generated by the funded project (e.g., toll fees). Bonds are widely used by U.S. cities and states and are beginning to be deployed by development banks in many countries.
- Use tax increment financing to finance improvements, which involves the city borrowing money for a civic project, such as constructing a transit station or park. The presence of this new amenity increases the property values of nearby homes and businesses, which in turn increases the amount of property taxes they pay each year (although the property tax rate is unchanged). The loan is paid off with the proceeds from the increased property tax revenues from the surrounding properties.
- Use proceeds from impact fees to fund urban mobility improvements. Many cities assess impact fees on developers to cover the costs of expanding services (e.g., fire protection, police protection, schools) to support new development. Also, a new development generates trips (more

people coming and going), which puts strain on the surrounding road and public transit networks. One of the best ways of mitigating traffic impacts is to use the impact fees to fund public transit, walking, and biking improvements. Portland encourages environmentally friendly buildings by giving breaks on impact fees to certified green buildings. Similar breaks might be offered for transit-oriented infill developments.

- Seek matching funds from regional or national government. Higher-level government agencies may offer funds to cover part of the cost of urban public transit projects.

- Enter a public–private partnership, where the private partner provides some funding in return for a benefit related to the project, such as exclusive rights in project operation for a period of time. For example, Denver's Eagle P3 project, an extension of the city's rail system, is being designed, built, financed, operated, and maintained through a partnership between the government and several private companies.[13]

- Levy a special sales or property tax. The tax is earmarked to fund one or more specific infrastructure projects and may include a sunset provision, which removes the tax once the project in question has been paid for. These sorts of taxes may require voter approval, which is often feasible to obtain if the benefits of the urban mobility improvements are clearly described. Note that this sort of tax calls a great deal of public attention to a project, which will then need to be defended in the media, so it is best used only for extremely large projects that will affect many residents (such as the initial construction or a major upgrade of a city's subway system).

- Make use of state infrastructure banks or revolving funds. Cities that are located in states with these finance options can obtain low-interest loans for infrastructure projects. The funds they pay back to the bank can be used later, by that same city or a different city, to fund more infrastructure projects.

- Rely on user fees. Bridges and similar limited-access roadways may charge tolls to help recover their construction costs or to fund transit. Most public transit systems charge fees for each ride, which can go toward paying for expansion of the transit system as well as daily operations and maintenance. Fees from parking and congestion pricing may be used to fund transit improvements that help alleviate congestion.

- Auction permits that allow construction at higher densities than otherwise permitted. This is a novel way to raise money that has been used most extensively in Brazil. Together, Rio de Janeiro and São Paulo have raised more than $3 billion using this mechanism.
- Facilitate corporate sponsorship of an infrastructure project. Sponsorship allows a corporation to contribute money to a project in return for rights to name some portion of that project after the corporation. Similarly, transit vehicles, stations, bicycle racks, and so on may bear advertisements that provide an ongoing funding stream for the city.

Political Opposition

Political opposition to new development or transportation infrastructure is common and can be an important barrier to overcome. Existing residents are one likely source of opposition. They may object to development for several reasons. A common reason is fear of losing the attributes of a city or town that they find valuable, such as a small-town atmosphere. Another reason is to increase the value of their properties: If little additional housing is built and demand for housing grows, property values rise. If a project would displace residents or businesses, residents who would be kicked out are likely to object.

Another source of opposition can be owners and operators of existing transit fleets. New infrastructure and transit projects can threaten their revenues, even if the projects are in the best interest of the city's residents.

From a city's perspective, the way to maintain its growth and prosperity in the future is to build new housing, commercial space, and transit systems in order to efficiently accommodate new residents, keep housing affordable, ease traffic, and generate tax revenues to fund city services. Additionally, a city or town has a responsibility to respect the interests of all residents, including renters, who are harmed by rising housing prices and may even be displaced.

The keys to reducing political opposition are to better manage residents' expectations about the way in which cities develop and the necessary consequences of economic prosperity and to prevent a small number of residents from undermining the will of the majority. Similarly, the concerns of existing transit owners and operators of transit should be tackled in any new transit plan.

The following approaches may help:

- Encourage developers to engage extensively with the local community early in the project development cycle to provide facts about their plans and build support. Local organizations such as environmental groups, the chamber of commerce, and public health groups have been known to support development that reduces car usage. Encourage developers to get supportive third-party groups on their side.
- Use polls to understand residents' views about a project rather than relying primarily on the results of city meetings, where residents who are vocal opponents and unlikely to reflect the views of most residents are more likely to appear.
- Zoning and building codes should be designed to enable high-density, transit-oriented, mixed-use development without causing individual projects to seek waivers (for use, height, setbacks, and so on). If no waiver is required, this reduces project risk and accelerates the timeline, thereby reducing developers' financing costs.
- Decisions on permits should be made based on the impacts of a development on the city as a whole, not only on the immediate area surrounding the development.
- Work with transit owners and operators to tackle their concerns. One way to do this is to use competitive concession bidding for new operations. Under competitive concession bidding, firms bid in offers to build new projects, with the government selecting the lowest-priced offer that meets the project's criteria. Another option is to fund programs that help retrain displaced workers or create new jobs.
- If a project would displace residents, engage these residents early to agree on payment and land for resettlement needs as part of a broader transit development package.

Case Studies

Guangzhou, China: Well-Designed Public Transit

Guangzhou, China has one of the highest-quality bus rapid transit systems in the world, as ranked by the Institute for Transportation & Development Policy (Figure 9-10).[14] Guangzhou's bus rapid transit system is the second largest in the world, behind only Bogotá's TransMilenio system, with daily ridership of up to 1 million people.[15] Operating since 2010, Guanzhou's system is the first [bus rapid transit] system to directly connect to a metro system and the first

Figure 9-10. Bus rapid transit in Guangzhou, China. (Photo by Minseong Kim [Own work], CC BY-SA 4.0, via Wikimedia Commons.)

[bus rapid transit] system in China to integrate bike parking into station design. Among other bus rapid transit sytems, it has the highest number of passenger boardings at stations, highest [bus rapid transit] bus frequency, and longest [transit] stations."[16] The system also uses smart cards for paying the fare, which allows passengers to transfer to other routes for free.[17]

Curitiba, Brazil: Mixed-Use and Transit-Oriented Development

Curitiba, Brazil (Figure 9-11) is known for its successful bus rapid transit system, and it is also a model of transit-oriented development, as a consequence of the city's deliberate and considered efforts to promote greater density along its transit corridors. Specific measures include:[18]

- Zoning all parcels within two blocks of the main bus rapid transit lines for mixed commercial and residential development, with generous height limits.
- Allowing property owners to sell their development rights in places where development was not wanted (such as in the historic city center), thereby protecting the historic character of those areas. Developers who purchased the development rights could build at higher densities elsewhere. The developers received inducements to use these rights to develop parcels located along the bus rapid transit corridors.
- Allowing developers of parcels near bus rapid transit lines to contribute to a city fund for public housing in return for the ability to build taller buildings.
- Restricting the construction of shopping centers to major transit corridors.

Copenhagen, Denmark: Facilitated Biking and Walking

Copenhagen is arguably the most bicycle-friendly city in the world, where 45 percent of residents commute to work or school via bicycle (Figure 9-12).[19] The city features 220 miles of dedicated cycle tracks, 14 miles of bike lanes, and 27 miles of off-street bike paths. Bicycles are allowed on trains, ferries, express buses, and taxis. Copenhagen was the first city to establish a bicycle-sharing program (in 1995), and the current iteration rents aluminum-framed bikes with GPS guidance. The high level of bicycling results in substantial social benefits, including reduced healthcare costs (from reduced air pollution, more exercise, and reduced car accidents) and reduced roadway congestion and maintenance.[20]

Copenhagen was also an early adopter of returning motor vehicle–dominated streets to pedestrian use. In 1962, the city closed a 1.1-kilometer (about two-thirds of a mile) length of the Strøget to cars.[21] Contrary to early fears, the plan was a success, and today, the Strøget is one of the longest pedestrian shopping areas in Europe.

London, UK: Car Control

In 2003, the city of London implemented a congestion pricing system (the "London congestion charge") for drivers operating vehicles within the central

(previous spread) **Figure 9-11.** Bus rapid transit lines in Curitiba, Brazil are bordered by a pattern of higher-density development. (Photo by Francisco Anzola [Flickr: Curitiba Centro], CC BY 2.0, via Wikimedia Commons.)

Figure 9-12. Mass bicycle traffic in Copenhagen, Denmark on special bike paths. (From goga18128/Shutterstock.com, Copenhagen, Denmark, 2016, Photograph, 2016.)

city, designated as the "charging zone," during weekdays (Figure 9-13). The zone is clearly denoted via signs and images painted on the roadway. Drivers may register vehicles online, paying for a single day, or register for automatic payment. Enforcement is fully automated: Cameras identify car license plates and a computerized system bills registered drivers for charges and issues penalty notices to unregistered drivers who enter the zone without paying. The charge is £11.50 per vehicle per day (discounted by 90 percent for residents and eliminated for various types of vehicles), and the penalty for driving without paying is £130.[22] Transport for London, the agency that operates the system, indicates

Figure 9-13. Congestion charging zone in London, United Kingdom. (By Mariordo [Mario Roberto Durán Ortiz] [Own work], CC BY-SA 3.0, via Wikimedia Commons.)

it has reduced traffic volumes by 10 percent relative to business-as-usual conditions and has raised more than £1 billion for public transport, road, bridge, walking, and cycling infrastructure projects.[23]

Conclusion

Getting urban form and mobility right is important. Smart urban design makes for a livable, dynamic city, whereas poor design can degrade quality of life, increase emissions, and lock in a development pattern whose consequences last decades or centuries.[24] Urban planners and policymakers can help guide a city's development through robust support for public transit coupled with mixed-use development; measures to facilitate biking and walking, including an interconnected street layout of small blocks; a focus on infill development; and car control measures. With the trend toward increased urbanization in the developed and especially in the developing world, smart urban design and mobility measures will be a crucial component of the transition to a clean energy future.

THE BUILDING SECTOR

The building sector is responsible for 8 percent of annual global greenhouse gas emissions today, with emissions of about 4 billion tons of CO_2.[1] Emissions are expected to grow to between 5 and 6 gigatons by 2050, and without additional policies the building sector will be responsible for 8 percent of cumulative emissions through 2050.[2] Buildings and appliances are also significant drivers for electricity demand (whose emissions are tackled in Section I, "The Power Sector"). For example, buildings are responsible for 54 percent of global electricity demand, and that share is expected to grow to nearly 60 percent by 2050.[3] When electricity emissions attributable to the building sector are included, its share of today's global greenhouse gas emissions increases to 20 percent and grows to 26 percent by 2050.[4] The growth in emissions is due largely to a growing building stock filled with more energy-consuming technologies.

Reducing emissions from the building sector requires improving the efficiency of building equipment, such as air conditioning and heating equipment, the thermal efficiency of buildings, and the efficiency of appliances used in buildings. Chapter 10 discusses how building codes and appliance standards can increase the efficiency of buildings and appliances.

Decarbonizing the building sector and reducing demand for electricity are an essential part of lowering our overall carbon emissions. Building codes and appliance standards can achieve at least 5 percent of the reductions required by 2050 (Figure S-3) and an even higher share in later years, because higher efficiency standards take years to reach full effect.

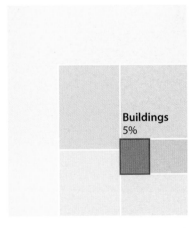

Buildings
5%

Figure S-3. Potential emission reductions from the building sector. (Analysis done using data with permission from the International Institute for Applied Systems Analysis [IIASA]. Data source: Tavoni et al., 2013. Data downloaded from the LIMITS Scenario database hosted at IIASA, https://tntcat.iiasa.ac.at/LIMITSPUBLICDB /dsd?Action=htmlpage&page=about.)

CHAPTER TEN

Building Codes and Appliance Standards

Residential and commercial buildings are major energy consumers, accounting for roughly 20 percent of delivered energy use and more than 50 percent of electricity use worldwide.[1] In terms of energy demand, buildings in urban areas, where a majority of the world's population lives, are particularly important. Furthermore, there is a continuing trend toward urbanization, particularly in the developing world: From 2014 to 2050, the population in urban areas is projected to increase from 3.88 billion to 6.34 billion,[2] with a corresponding growth in building floor space. Not only is the world urbanizing; it is also developing. As people rise out of poverty, they will demand more energy-consuming services, such as lighting, air conditioning, and television. As the number of city dwellers grows and they demand more services, there is potential for substantial increases in building energy use. The International Energy Agency projects global building energy use to grow at about 1.3 percent per year through 2040.[3] Smart, ambitious policy can drive efficiency increases in building components, helping to decouple demand for energy from demand for building services. This decoupling is a key part of the transition to a low-carbon future.

Most buildings and building equipment (e.g., furnaces and air conditioners) last decades. Rapidly urbanizing countries with high rates of new building construction, such as China, India, and Nigeria,[4] are in a crucial period during which good building efficiency policies can have an outsize impact. Buildings constructed today will lock in energy use patterns for many years to come. This speaks not only to the importance but also to the urgency of policy to ensure that new and renovated buildings are well built and efficient. Two of the most effective policies to achieve these goals are building codes

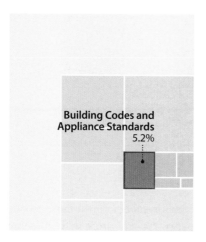

Figure 10-1. Potential emission reductions from building codes and appliance standards. (Analysis done using data with permission from the International Institute for Applied Systems Analysis [IIASA]. Data source: Tavoni et al., 2013. Data downloaded from the LIMITS Scenario database hosted at IIASA, https://tntcat.iiasa .ac.at/LIMITSPUBLICDB/dsd?Action=htmlpage& page=about.)

and appliance standards. Together they can achieve at least 5 percent of the reductions needed by 2050 to achieve the two-degree target (Figure 10-1).

Policy Description and Goal

Building Codes

Building codes are regulations that impose requirements on the design and construction of buildings. Codes can serve a variety of social purposes, such as protecting the health and safety of occupants and reducing impacts on nearby public spaces. However, when the goal is to limit energy use and emissions, building codes typically set minimum technology requirements or performance criteria that new or modified buildings must meet. For example, building codes may specify how effectively windows or walls must insulate the interior from exterior temperatures or set a required efficiency level for central heating or air conditioning systems.

In contrast, whole building energy performance standards specify a maximum energy use for the entire building rather than for individual components of that building. These standards may be used for code compliance, in voluntary certification programs, to determine eligibility for tax incentives, and for other purposes.[5]

Appliance Standards

Similar to energy efficiency building codes, appliance standards set minimum energy performance requirements for new appliances. Appliance standards typically specify the maximum amount of energy that an appliance of a given capacity and type may use in order to perform a particular service. An appliance standard might mandate that every specific model comply with the standard, or it might allow manufacturers to meet the standard via a sales-weighted average of the appliances of a given type that they sell.

An important distinction between building codes and appliance standards relates to monitoring and enforcement. Most buildings are one-of-a-kind or, at most, are built in small numbers and are site-specific. Buildings "are the largest handmade objects in the economy."[6] Accordingly, inspections and energy audits of every individual building may be necessary to determine whether those buildings comply with a performance standard. In contrast, large numbers of identical appliances are produced in factories, and their energy efficiency can be tested "upstream" on just a few units provided by a manufacturer or importer during a verification process. Once in wide distribution, appliance standards also require ongoing enforcement to ensure manufacturers aren't cheating standards. For example, check testing, where for-sale products are periodically randomly checked and tested, and import controls can help uncover cheating by manufacturers. However, appliance standards are more suited to initial verification than building codes, so they are often easier for lower-income regions to implement. In both instances follow-up enforcement is important.

When to Apply This Policy

A key element of both building codes and appliance standards is their ability to mandate performance improvements that are economically beneficial to consumers yet for a number of reasons fail to be adopted in the market. Because of the characteristics of the building sector, market failures of various sorts occur almost everywhere, so building codes have very broad applicability. One common example is a misalignment of incentives between those who pay for energy and those who pay for buildings and appliances. Many households and most businesses do not own the buildings they occupy; rather, they are tenants. Property owners lack an economic incentive to improve the energy efficiency of the building because the tenants are generally responsible for paying the utility bills. (Information barriers and other issues pertaining to human psychology make it difficult to price energy efficiency upgrades into the rent charged to tenants.[7]) Meanwhile, tenants may be prohibited from making major changes to the building. Even if allowed, tenants may balk at putting money into buildings that they do not own, particularly if they believe they might move before the capital cost is recouped via energy savings. Building codes and appliance standards set a minimum level of performance that building owners are expected to provide, thereby ensuring that the energy savings (which are larger than the capital cost) are realized.

Examples of other market failures that are partially mitigated by building codes and appliance standards include consumers' short or inconsistent discount rates,[8] insensitivity to rebate or subsidy policies, imperfect information, and high transaction costs.[9] Amory Lovins provides a more detailed discussion of various engineering and institutional reasons for inefficient building design and construction.[10]

Efficiency improvements do not necessarily increase the cost of equipment relative to less efficient models, because the policy can spur companies to engage in research and development (R&D) that lowers the cost of efficiency improvements over time. (Thus, the fact that a more efficient appliance may be more expensive than a less efficient appliance today is not a good guide to whether a performance standard would result in increased appliance costs in future years.) For example, Figure 10-2 shows that refrigerator efficiency in the United States has improved since 1972, while actual consumer prices have declined and refrigerated volume has increased.[11] These efficiency improvements were driven in part by a series of standards, beginning in 1978.[12]

Building codes and appliance standards can and should be designed to improve over time, thereby providing a clear signal to manufacturers that R&D investments in efficiency will pay off. When the standard comes into effect, a company that invested in R&D can produce better or cheaper appliances and building components than are available from competitors. This allows the manufacturers to gain market share.[13] If all manufacturers engage in R&D, this drives down prices across the board and benefits consumers.

In contrast, policies that provide a rebate for the most efficient products on the market tend to encourage individuals and businesses to purchase these products, but they often do not encourage manufacturers to engage in R&D and improve their products' performance over time (because the rebate qualification thresholds typically are based on the range of products already available in the market). Product labels, on the other hand, have been shown to motivate manufacturers to engage in R&D and meet increasing efficiency requirements.

Another advantage of building codes and appliance standards is their low cost to implement. A rebate or subsidy for efficient products may cost government a great deal of money because it must pay some fraction of every rebate-qualifying product that is sold.[14] A tax on inefficient products or especially a

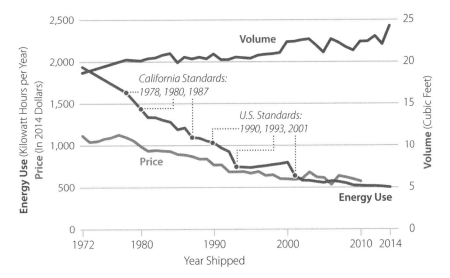

Figure 10-2. U.S. refrigerators have become less expensive while becoming more efficient and larger. ("Annual Energy Use, Volume and Real Price of New Refrigerators," U.S. Department of Energy, 2014, https://energy.gov/sites/prod/files/2014/08/f18/energy_use_new_refrigerator_chart.pdf.)

tax on fuel may cost building owners or occupants a significant amount of money. A standard does not require large payments between government and energy users throughout society. The only government costs relate to the initial establishment of standards (e.g., technical analysis of products, meeting with industry) and the monitoring and enforcement of the standards. These costs may be lower for appliances, where the standard can be enforced upstream, at a limited number of points of manufacture or importation. Building component standards are harder to enforce because they may involve inspections of individual buildings, but nonetheless, they are cheaper than many alternative policy options. Additionally, the cost of codes and standards can be scaled by varying the extent and type of analysis and stakeholder engagement. Greater resources will yield more rigorous standards, but even without this level of rigor, standards can be effective.

Another benefit of standards for appliances and some types of building components is the ability to cause efficiency improvements in areas beyond the control of the policymaker. For example, if a large region such as California implements stringent efficiency standards for televisions, manufacturers that want to sell televisions in California must produce standard-compliant models that will appeal to consumers. They may be obligated to conduct R&D in order to maintain their ability to sell attractively priced and functional televisions in California in the future. They also must retool their assembly lines to produce these improved televisions. Having already made these R&D and manufacturing investments, the appliance manufacturer may determine that it is more cost-effective simply to sell the standard-compliant televisions throughout the United States rather than maintain production of less efficient televisions that can be sold only in some states. The larger the number of regions that enact standards, the more likely manufacturers are to choose to comply with those standards universally rather than only in the jurisdictions where the standards are in force.

Detailed Design Recommendations

Policy Design Principles
The following policy design principles apply to building codes and appliance standards.

Create Long-Term Certainty to Provide Businesses with a Fair Planning Horizon
It may take years for manufacturers of building systems and appliances to make the necessary R&D investments and changes in their manufacturing lines that reduce energy consumption without adding greatly to the building's cost and without compromising other performance characteristics. If standards are set just a few years at a time, businesses may be unsure whether standards will continue to become more stringent in future years. This means they will not know whether modest investment in near-term improvements is sufficient or whether a larger investment in a new technology or major design change might be worthwhile.

Because the process of selecting a building site, acquiring building permits, obtaining financing, and constructing a building can stretch for many years, it is particularly important for architects, engineers, and financiers to know building code requirements over a sufficiently long time horizon.

Standards can help companies that manufacture appliances and building systems justify investments in R&D and energy efficiency to their shareholders, who otherwise might be skeptical of the value of these expenditures.

Companies that make a serious commitment to R&D may even see this as their competitive advantage.

Build In Continuous Improvement

Building codes and appliance standards are valuable because they increase efficiency of the building stock and eliminate the worst-performing buildings and products over time. If the standards are allowed to stagnate, they are not serving their primary purpose. For example, the U.S. Government Accountability Office estimates that failure to update appliance standards in the United States in the 1990s and early 2000s will cost consumers at least $28 billion through 2030.[15] Forgone energy savings imply further damages, such as greater emissions of pollutants that cause premature deaths and drive climate change.

Focus Standards on Outcomes, Not Technologies

Standards should set energy performance targets rather than specifying particular technologies to be used. This allows architects and engineers the maximum flexibility in designing buildings and products that achieve their financial, aesthetic, and functional goals while achieving the requisite energy performance. Buildings are diverse and can have features that are site-specific and are particular to the building's intended purpose, making prescription of specific technologies unwise. The same air handling, heating, or envelope components that work well in an office building may be ill suited for a residential building or a warehouse. For example, to reduce lighting energy consumption, extensive use of skylights, windows, other openings, or reflective surfaces (collectively called daylighting techniques) may be suitable for a building on a site with sufficient sunshine and intended to be used primarily in the daytime, but these techniques may not achieve the desired energy savings in a building on a different site or intended to be used at night.

Prevent Gaming via Simplicity and Avoiding Loopholes

Standards should be written with simplicity and clarity to prevent manufacturers and building designers from gaming the standards. Standards should be written in consultation with a diversity of experts and stakeholders and carefully checked by government scientists and other experts for possible loopholes. In general, specifying overall energy performance targets with as little variance as possible, given building characteristics, will tend to minimize loopholes, although the resulting standard may be more difficult for some building types to comply with than others.

Additional Design Considerations

If building and appliance performance standards are designed according to the principles just described, there are only a few remaining points to consider.

Require Thorough Monitoring and Enforcement

To be effective, building codes and appliance standards must be subject to rigorous testing and enforcement. Local governments must have adequate funding to properly review building plans and to hire, train, and dispatch building inspectors. Building inspectors should be assigned randomly to each building or development, to avoid corruption, and for big projects, inspections should be alternated between two or more inspectors. Similarly, appliances must be subject to random, standardized testing to ensure they perform in accordance with an appliance standard. Where enforcement is weak, manufacturers and building owners will often fail to meet the standards. For example, in China, where enforcement staff and funding are inadequate, third-party testing has demonstrated that many appliances fail to meet national standards, even though manufacturers self-report their appliances as up to code.[16] Similarly, China worked to develop an improved building inspection system in the late 2000s. Before this system, just 21 percent of medium and large construction projects in urban areas complied with building codes. The improved inspection system greatly increased compliance rates of these types of projects. However, there are almost no inspections of rural developments, nor of small developments in urban areas, so compliance rates among these building types are unknown (and probably low).[17] To avoid cheating by manufacturers or importers of appliances, the government should use a randomized process to purchase appliances to be tested for energy efficiency rather than allowing the manufacturer or importer to submit a unit for testing. The government should conduct spot tests, to complement manufacturers' certifying that their own products comply with standards. In regions with strong economic and legal infrastructure, these tests can be occasional, happening only every few years. But in regions with weaker infrastructure, government should test more frequently.

Ensure Sufficient Knowledge of Green Building Design and Construction

Buildings are almost always designed and built by hand, not mass produced in factories. The ability of a building to meet energy efficiency standards depends in part on the ability of that building's designer to anticipate and correctly design energy-saving features and for the construction companies and contractors erecting the building to use best practices. Small matters such as proper air sealing of ducts can make a substantial difference in final energy use. Construction firms must be attentive to energy-efficient construction and must build structures of high quality. Therefore, they must have the necessary knowledge and capabilities.

Government can help by providing or funding education and training programs for the building industry and leading by example through performance-based building procurement, in which procurement contracts for new buildings specify minimum energy performance criteria.[18] Government can also ensure that energy use is monitored, and buildings that fail to meet the energy efficiency standard should be penalized and the defects immediately corrected at the owner's expense. This gives the building owner an incentive to ensure the building achieves promised energy savings. The building owner can then select the firms with the best reputations for energy-efficient building practices and include energy performance requirements in construction contracts.

Couple Codes and Standards with Strong Complementary Programs

Building codes and appliance standards rely on stock turnover to achieve energy savings. Appliances often take more than 10 years to reach the end of their lifetimes, and some building components, particularly parts of the building envelope, can last 50 years or more. Therefore, it may be many years before the full benefit of building codes and appliance standards is realized. To help mitigate this problem, building codes and appliance standards should be coupled with strong retrofitting programs (also known as retrocommissioning), including subsidies or other incentives, to help accelerate stock turnover. Applying strong building codes in places that are rapidly urbanizing, such as India, China, and Nigeria, can affect buildings when they are first built rather than waiting for building turnover, but rapidly developing and urbanizing areas are also some of the places with the greatest enforcement challenges, so codes may be a higher-risk, higher-reward policy option in these places.

Building codes and appliance standards generally affect only the bottom of the market, removing the worst-performing products each year. They do not provide incentives for people to purchase the top-performing products instead of products that are minimally compliant with the codes. This drawback can be remedied by pairing codes with other policies that promote uptake of the top-performing products, such as clear labels that show energy savings, subsidies or rebates for the most efficient products, and carbon pricing that raises the costs of owning inefficient products.

Financing Efficiency Upgrades

Access to low-cost capital can greatly improve the cost-effectiveness of retrofits and appliance upgrades and encourage turnover to more efficient components before the end of their useful lives, which can be 50 years or more. However, building owners, particularly low-income people, may lack the upfront capital and credit-worthiness needed to pay for efficiency upgrades. Energy service companies (ESCOs), property-assessed clean energy (PACE) financing, on-bill repayment, and green bank financing are a few successful policies that overcome financing barriers.

ESCOs operate as financial intermediaries between customers and project lenders, often utilities. ESCOs propose and manage a suite of efficiency investments and use energy bill savings to cover the cost of the project and repay the financier. However, because ESCOs retain some of the revenue that results from lower energy bills, and because of the administrative costs of an intermediary, the ESCO model can reduce the number of investable projects.[19]

PACE financing is a tool under which city governments grant loans to building owners to purchase and install energy efficiency measures, which are paid back through property tax bills, often over 15 to 20 years. Because PACE financing allows small building owners to affordably finance the installation of efficiency measures, it can increase the penetration of energy-efficient appliances in small buildings, which has often been difficult to achieve. However, PACE financing requires municipal bonds to finance energy efficiency measures and may be hard to implement in regions without credit-worthy municipalities.[20]

Under on-bill repayment, a utility pays for the upfront costs of efficiency measures and collects those costs, plus reasonable returns, over time via customer utility bills. Customers are able to offset these costs with lower energy bills from efficiency measures. The financing costs "stay with the meter," meaning the charge remains whether or not the building changes hands, making the financing straightforward compared with PACE financing. Depending on how the program is structured, as a loan or merely an additional charge to the bill, credit-worthiness of the building tenant may not matter, increasing access for low-income customers.[21]

Green banks provide low-interest financing to clean energy and efficiency developers, backed by government funds. Efficiency portfolios tend to be stable investments delivering steady cash flows over time, but finding a financier that is willing to take on small or complex efficiency projects can be difficult. As a result, green banks have emerged as a leading lender for medium- and large-scale clean energy and efficiency projects, providing a promising option for financing projects in the future.[22]

Case Studies

California's Energy Code

Title 24 of California's Energy Code was established in 1978 to provide a single coordinated, comprehensive building energy standard that governs the design and construction of all California buildings, associated facilities, and equipment. In particular, Part 6 of the California Energy Code contains requirements for building energy use and energy efficiency. The standards require a minimum energy efficiency level and provide buildings with two options for meeting the standard by using either performance-based or prescriptive methods.[23] The code also offers voluntary "Reach Standards" for buildings that strive to achieve additional energy savings.[24] The California Energy Commission has approved a handful of energy analysis computer programs, including both public and private domain software, to test for building compliance.[25] A building permit is required to construct or substantially alter a structure. To receive a permit, builders must submit project plans to city officials and be in compliance with Title 24 and other regulations. The California Energy Commission has a "Building Permit Violation Referral form" that anyone can use to report unpermitted construction activity.[26]

Title 24 requirements are updated by the California Energy Commission on a 3-year cycle.[27] Six California utilities operate a Codes and Standards Enhancement Initiative, which supports the standard revision process through public meetings and research to help identify areas that show promise for energy savings potential. For the upcoming Title 24 revision, the Codes and Standards Enhancement Initiative provides guidance on improvements in lighting, HVAC, indoor air quality, building envelope, water heating, and demand response.[28] Members of the public may also propose measures for consideration.[29] This update process helps ensure that Title 24 requirements do not become stagnant and that tighter standards remain achievable in commercial practice.

Title 24 has saved more than $74 billion in electricity bills and helped keep California's per capita electricity usage stable over the last four decades,[30] even though state gross domestic product (after adjustment for inflation) more than tripled during this period.[31] The California Energy Commission has found that since 1978, the standards have avoided "more than 250 million metric tons of greenhouse gas emissions (the equivalent of removing 37 million cars off California roads)."[32] It estimates that the newest, 2016 revision of the Title

24 standards will save a typical residence $7,400 over the course of a 30-year mortgage while increasing the initial price of the residence by only $2,700.[33]

Mexico's National Energy Efficiency Standards

Mexico is a leader among developing countries in building and appliance energy efficiency standards and could serve as a model for other nations that want to implement or strengthen standards.

Mexico's first mandatory standards came into force in 1994 for residential refrigerators, air conditioners, and three-phase electric motors, all important electricity-consuming devices in Mexico.[34] Coverage has since expanded to 29 types of appliances, equipment, and building components.[35] The law establishing Mexico's standards requires them to be updated every 5 years,[36] helping to avoid stagnation.

Mexico intentionally selected the stringency of its standards and designed its test procedures to mirror those used in the United States. This harmonization lowered barriers to the trade in these appliances and expanded the market for Mexican manufacturers. Mexican exports of refrigerators to the United States "increased 9-fold from 401 million U.S. dollars to about 3.7 billion U.S. dollars" per year from 2000 to 2014,[37] whereas the domestic Mexican market only doubled during that period.

The implementation of energy efficiency standards did not lead to higher prices for consumers. Figure 10-3 shows that the domestic Mexican market for combination refrigerator/freezers, refrigerators, split air conditioners, and window air conditioners grew considerably over the 1999 to 2013 period while per-unit prices of all these technologies dropped (except for window air conditioners, which remained flat). During this period, the devices' features increased while durability and quality were not harmed. The most likely explanation for the price declines is that Mexican manufacturers and importers were able to take advantage of economies of scale and adapt successfully to more stringent efficiency requirements while keeping prices down.[38]

The standards also led to reductions in overall electricity use and in peak-time demand (Figure 10-4), reducing costs to society and lowering greenhouse gas and conventional pollutant emissions.

Mexico eased implementation of its standards by involving many stakeholders. Representatives from the private sector, equipment manufacturers, academia, the government, and nongovernment organizations all participate in an advisory committee that guides the standard-setting process.[39] Industry

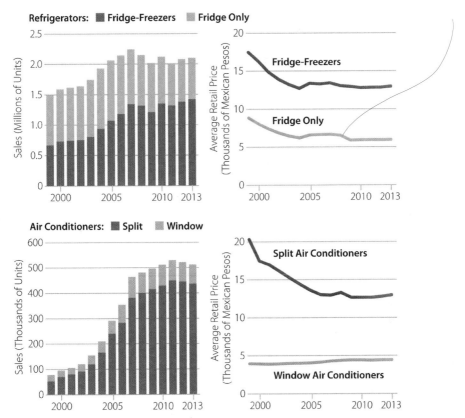

Figure 10-3. Mexican refrigerators and air conditioners have become cheaper while becoming more efficient. (Graphs reproduced with permission from Lawrence Berkeley National Laboratory, McNeil et al.)

representatives have praised the standards for allowing them to compete on a fair playing field and for signaling that investments in improved manufacturing processes and technology would be rewarded.[40]

The standards have not been without downsides for Mexican manufacturers. The same improvements in energy efficiency that opened up the U.S. market to Mexican products have caused those products to be less competitive in developing Central American nations that lack energy efficiency standards. Consumers in these nations can buy inefficient, Asian-made devices more cheaply than Mexican equipment, and as a result, Mexican exports of

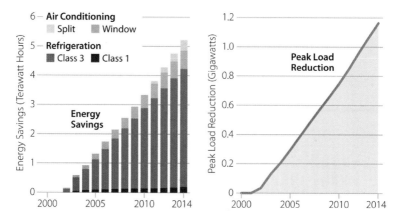

Figure 10-4. Energy efficiency standards for Mexican refrigerators and air conditioners have reduced total and peak energy use. (Graph data reproduced with permission from Lawrence Berkeley National Laboratory, McNeil et al.)

equipment to Central American nations have dropped.[41] This holds a lesson for developing countries that manufacture appliances for export: It may be wise to harmonize standards with a large export market to offset any drop-off in sales to developing markets that lack efficiency standards. Alternatively, groups of economically interrelated developing countries may consider enacting standards all at once, as a regional coalition, to prevent the loss of market share to inefficient imports from outside the coalition.

Conclusion

Building codes and appliance standards are a key policy for reducing emissions from the building sector, which is particularly important in rapidly urbanizing countries. Standards are effective at overcoming various market barriers that are common in the building sector and limit the effectiveness of economic incentive policies. When properly implemented, standards drive down emissions year after year while achieving net savings for society. The best standards in the world are technology finding, are publicly known years in advance, are resistant to gaming, and have built-in mechanisms to tighten the standards over time. Proper monitoring and enforcement of building codes and appliance standards can be a particular challenge. However, when well written and enforced, building codes are a crucial way to save energy, reduce emissions, and save money—steps on the path to a clean energy future.

THE INDUSTRY SECTOR

The industry sector, including agriculture and waste, is responsible for 38 percent of annual global greenhouse gas emissions today, with CO_2e emissions of about 19 billion tons.[1] Emissions are expected to grow to more than 42 billion tons by 2050.[2] Without additional policies, the industry sector will be responsible for 49 percent of cumulative emissions through 2050.[3] The industry sector is also a significant driver of electricity demand (whose emissions are tackled in Section I, "The Power Sector"). For example, the industry sector is responsible for roughly 44 percent of global electricity demand, although that share is expected to fall to about 36 percent by 2050.[4]

Industry sector emissions can be broken into two categories: emissions from fossil fuel combustion for energy use and process emissions. Process emissions are emissions released in industrial processes such as cement clinker manufacture and metallurgical coal coking. Additionally, we categorize nonenergy emissions in agriculture and waste as process emissions. The share of industrial emissions from processes is significant. At least 10 billion tons of CO_2e per year are from industrial processes: about 5.2 billion tons of CO_2e per year from agriculture, 1.5 billion tons from waste, and 3.2 billion tons from more traditional manufacturing-related processes.[5]

Reducing emissions from the industry sector therefore requires both improving the efficiency of industrial production, thus lowering demand for energy, and eliminating emissions from industrial processes. Chapter 11 discusses how to improve efficiency in industry. Chapter 12 discusses the sources of industrial process emissions and how to reduce them.

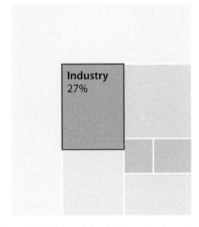

Figure S-4. Potential emission reductions from the industry sector. (Analysis done using data with permission from the International Institute for Applied Systems Analysis [IIASA]. Data source: Tavoni et al., 2013. Data downloaded from the LIMITS Scenario database hosted at IIASA, https://tntcat.iiasa.ac.at/LIMITSPUBLICDB /dsd?Action=htmlpage&page=about.)

Heavily decarbonizing the industry sector is essential to hitting our climate targets. Improvements in industrial energy efficiency can achieve 16 percent of the necessary reductions by 2050, and reducing process emissions can achieve at least 10 percent of the necessary reductions (Figure S-4).[6]

CHAPTER ELEVEN

Industrial Energy Efficiency

Industry plays a central role in the world's economy and is responsible for more than 40 percent of global energy consumption, more than any other sector.[1,2] In countries whose economies are heavily based on manufacturing, this percentage is higher. For example, in China, industry consumed 69 percent of total energy in 2014.[3]

Because of the central role industry plays in driving energy use and pollutant emissions, careful consideration of policies to reduce industrial energy use is warranted. However, policies targeting industrial energy use are less frequently discussed than policies affecting the transportation or electricity generation sectors. This may be due to a lack of familiarity with industrial energy efficiency opportunities and technologies or the incorrect[4] assumption that industrial facility owners are already undertaking all available, cost-effective means to reduce energy use. In fact, many technologies exist that can lower energy use—often while saving money in the long run—and there remain tremendous opportunities for efficiency improvement, typically at a lower cost than improvements targeting at other sectors.[5] Together, industrial energy efficiency policies can achieve at least 16 percent of the greenhouse gas emission reductions needed to hit the two-degree target (Figure 11-1).

Policy Description and Goal

Eight key types of policies, used in combination, can efficiently reduce industrial energy use and accelerate the transition to a clean energy economy: education and technical assistance, financing, financial incentives, mandatory targets, equipment standards, energy management system promotion, energy audit and energy manager requirements, and policies to support research and development (R&D). These policies help industry invest in higher-quality equipment that will save energy and quickly pay for itself, increasing the long-term competitiveness of affected businesses.

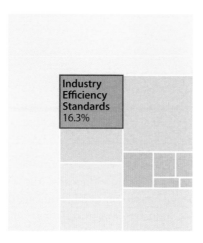

Figure 11-1. Potential emission reductions from industrial energy efficiency policies. (Analysis done using data with permission from the International Institute for Applied Systems Analysis [IIASA]. Data source: Tavoni et al., 2013. Data downloaded from the LIMITS Scenario database hosted at IIASA, https://tntcat.iiasa .ac.at/LIMITSPUBLICDB/dsd?Action=htmlpage &page=about.)

Industrial Energy Efficiency Technologies and Measures

Although this chapter discusses primarily policies and programs to promote industrial energy efficiency, it is worthwhile to review some of the major mechanisms by which energy savings in industry can be achieved. Without this knowledge, it can be unclear what steps the policies are designed to encourage, hampering efforts to understand good policy design.

Although industries vary in the products they create, similar equipment and processes are used in multiple industries. For example, every industry has electric motors, most have pumps, and all have heating or cooling demands. Measures may need to be adapted to fit the specific conditions or requirements at a given facility, but some measures that can be effective for a variety of industrial facilities include those discussed in this section.

Waste Heat Recovery

In the United States, between 20 and 50 percent of the energy used by industry "is lost as waste heat in the form of hot exhaust gases, cooling water, and heat lost from hot equipment surfaces and heated products."[6] Heat from high-temperature exhaust and water can be recovered and used in a variety of ways. It may be used to preheat loads (materials entering the system, such as combustion air or feedwater going into boilers), so that less fuel is needed to raise the temperature of these inputs once inside the system.[7] Another common application is to use the heat to drive an electric generator, producing electricity for use by the facility.[8] A facility that uses waste heat to generate electricity is sometimes called a combined heat and power or cogeneration facility. It is even possible to use waste heat for cooling purposes by adding an absorption chiller, a device that uses heat to drive a refrigeration cycle.[9]

Properly Sized and Variable-Speed Motors

Motors are used for a variety of purposes in industrial facilities, such as moving materials, running assembly lines, and controlling equipment. The International Energy Agency reports that a full 40 percent of all electricity use is in motors and that a quarter of this at least can be saved, reducing global electricity demand by 10 percent.[10] Several techniques can be used to reduce motor energy use. A motor, fan, or pump that is larger and more powerful than necessary wastes energy.[11] Ensuring that overly large equipment is not purchased can reduce the purchase price of the motors while saving energy. Motors must be able to accommodate their peak loads, so designing the industrial process to lower the peak load on a single motor (e.g., by evening out material flows between components or across time) may enable motors to be replaced with smaller, more efficient models.

A motor may be equipped with a variable-speed drive, control equipment that regulates the amount of electricity fed to the motor to adjust the motor's speed and torque output. Variable-speed drives save energy in fan, pump, and certain other applications by ensuring the motor consumes only as much power as is necessary to accomplish its task. Fewer than 10 percent of motors globally are controlled by variable-speed drives, so there remain large opportunities for energy savings.[12] (The alternative—controlling motor output with a valve, for example—is analogous to driving a car with a foot all the way down on the gas while regulating speed with the brakes.)

High-Efficiency Compressed Air Systems and Alternatives

Compressed air is used in industrial facilities for tasks such as cooling, agitating, or mixing substances; operating pneumatic cylinders; inflating packages; cleaning parts; and removing debris. Compressed air systems inherently have low efficiency, and the best option is sometimes to replace the compressed air system with another mechanism to accomplish the same task. For instance, fans or blowers, motors, vacuum pumps, and brushes may be substituted for compressed air in some cases.[13] When compressed air must be used, efficiency can be improved by frequently checking filters and clearing blockages; minimizing air leaks; using multiple, small compressors with sequencing controls rather than a larger compressor that always runs; and keeping the system at the lowest possible air pressure (using devices such as a booster or cylinder bore to increase air pressure locally for specific applications that require higher pressures).[14]

High-Efficiency, Properly Sized, and Condensing Boilers

Boilers are used in industrial facilities to generate steam, which is used for a variety of purposes, such as to turn turbines or to heat kilns in cement plants. Apart from waste heat recovery (discussed earlier), a variety of steps may be taken to improve boiler efficiency. Process control technologies can monitor the exhaust stream and optimize the air–fuel mixture entering the boiler. Reduction of air leaks and excess air saves energy, as more of the energy goes into generating steam than heating the air. Ensuring boilers are not larger than needed, improving boiler insulation with modern materials, and regularly removing fouling or scaling on part surfaces all improve energy efficiency.[15] Condensing boilers extract energy from the steam produced in combustion, thereby improving the efficiency of the system as a whole. Also, multiple boilers can be used in series, with low loads handled by one boiler and additional boilers fired up only when loads are higher.

Building Upgrades (Lighting and HVAC Systems)

Industrial firms can achieve energy savings by upgrading the lighting and heating, ventilation, and air conditioning (HVAC) systems of their buildings. Lighting efficiency can be improved by switching to more efficient types of bulbs (such as LEDs), using lighting control systems that illuminate areas only where light is needed, and using more natural light. HVAC efficiency can be improved via building or duct insulation and air sealing, upgrading to more efficient HVAC equipment, and using temperature setbacks during nonworking hours. Industrial automation provides a unique opportunity to reduce lighting and HVAC energy use because building services are needed only when human workers are present. Japanese robot manufacturer FANUC, a pioneer in factory automation, has a factory that can run for up to 30 days at a time with no human intervention. "'Not only is it lights-out,' says FANUC vice president Gary Zywiol, 'we turn off the air conditioning and heat too.'"[16]

Cog Belts

Belts are used in industrial processes to transfer rotational motion between components. "Cog belts" (those with teeth molded into their inner diameters) are more efficient, run cooler, and last longer than traditional belts with a smooth lower surface.[17]

Smart Monitoring and Controls

Industrial facilities may use computer hardware, software, and sensors that monitor and optimize the energy use of building systems and industrial equipment. These systems can help plant operators quickly detect energy waste, such as when devices are consuming more energy than expected, consuming energy while in standby mode, or in need of maintenance.[18] Newer "smart" controls may use sophisticated learning algorithms, achieving even greater energy savings.

Energy Management Systems

In addition to smart monitoring and controls, industries may use energy management systems. An energy management system is not a particular technology but rather an internal governance system or processes that companies follow in order to "systematically track, analyze, and reduce energy demand."[19] The most widely known guideline for energy management systems is ISO 50001, a set of requirements established by the International Organization for Standardization that include an energy planning process establishing baseline energy use, identifying energy performance indicators, setting objectives or targets, forming action plans, and conducting periodic measurement and internal audits.[20] Adopting an energy management system helps ensure that energy savings are not overlooked amid the variety of other goals that companies seek to achieve through their operations and investment.

Design for a Circular Economy

The term *circular economy* refers to the idea of reusing products and materials for the highest or best use for which they are suitable once the first user is done with the product. Industry can help with each stage in this process.[21] For example:

- A product that is still in good working order could be passed on to another user. Industry can help make this possible by designing products for durability and longevity.
- A product that is broken may be repaired rather than replaced. Industry can assist by ensuring the product can be opened and its workings

accessed, by making parts removable, and by using standard-compliant parts or making interchangeable parts available.

- A product that is too worn or damaged to be repaired could be sent back to the producer for remanufacturing or refurbishment. Industry can establish takeback or buyback programs for older devices.
- A product that is too damaged or too outdated to be refurbished can be recycled. Industry can help by designing products for recyclability and by making new products out of recycled materials.

Policies to Promote Industrial Energy Efficiency

Policymakers can take a variety of steps to encourage or require industry to adopt stronger energy efficiency measures.

Equipment Standards

Many types of equipment are common to industries and can be governed by efficiency standards in a way similar to building component or appliance standards. Standards specify a maximum allowable energy use or minimum efficiency that equipment must achieve. For example, the U.S. Department of Energy has specified mandatory energy efficiency standards for electric motors, pumps, commercial boilers, and various other types of equipment used in industrial facilities.[22] Appropriate standards may be complex because of the various types of equipment available. For example, the United States has set standards at different levels for three types of electric motors, divided into up to 25 horsepower categories, further divided by pole configuration and whether the motor is open or enclosed.[23] Equipment standards should be applied to both domestically produced and imported equipment.

Education and Technical Assistance

Government may provide education about energy efficiency options and technical assistance in their implementation. This role is best suited for a regulatory agency that possesses deep expertise on energy efficiency measures. An agency may target outreach to industries or companies that would particularly benefit, such as those that have large energy savings potential or those that are in need of education and assistance. Voluntary programs must emphasize cost-effective energy savings–those with reasonable payback periods–and should allow upgrades to be made in accordance with industrial facilities' operational schedules and capital investment cycles. Energy savings should be measured and verified, to assure the industries that participation in the program delivered the expected savings and to assure the public that the government program is reducing energy consumption and emissions.[24]

Education and technical assistance may be particularly important in developing countries where technical expertise is in short supply. For example, the Institute for Industrial Productivity determined that foundries in India suffer "from technological obsolescence, poor management practices and paucity of funds." Training local people in how to conduct energy audits and meeting with foundry owners were key parts of the institute's strategy to help the foundry industry improve its methods.[25]

Financing

Many industrial energy efficiency measures require upfront expenditures to purchase or upgrade capital equipment. Even if the payback period on this investment is reasonable, it can be difficult for some companies to set aside the money for upgrades. Government can make this easier by offering access to low-cost financing. For example, a government can establish a fund that is used to make low-interest loans to industries for energy efficiency retrofits and upgrades. As companies pay back their loans, this replenishes the fund, which can then be used to make loans to new companies. A government may offer these loans directly or might provide money to commercial banks and allow them to administer the loans.[26]

A similar approach is to establish a green bank, a publicly funded organization that attempts to use public money to leverage private investment. For example, a green bank might partner with a commercial lender to supply money for an efficiency project or might offer better interest rates and access to more credit for a borrower who otherwise would not qualify for a commercial loan.[27]

A government might consider getting initial capital to fund industrial energy efficiency programs, such as a fund or green bank, by issuing green bonds for the purpose. Green bonds are similar to other bonds, but their use is earmarked for specific, environmentally beneficial projects.[28]

Many companies may lack expertise in how to select, install, and monitor efficiency upgrades. To help them, government can direct companies that receive financing to work with energy service companies (ESCOs), businesses that design and implement energy-saving projects for clients. ESCOs can handle many of the technical aspects of the upgrade projects for financing recipients. This may simplify the upgrade process for participating companies.

Financial Incentives

Often it can be difficult for industry to secure the upfront capital needed to make energy efficiency upgrades, even if those upgrades would pay for themselves through energy savings. For example, in many cases companies separate capital budgets from operating budgets, and because energy savings accrue in operating budgets but efficiency upgrades are paid for through capital budgets, industry often finds it hard to make a good accounting case for investing in efficiency upgrades. Financial incentives can help overcome this and other barriers, such as choosing between two equally attractive investment options.

Governments may use financial rewards or penalties to encourage adoption of energy efficiency measures. For example, in China electric utilities (which are government owned) charge higher electricity rates to companies with higher electricity use per unit of product produced, improving the cost-effectiveness of reducing electricity consumption per unit of industrial output. Similarly, financial institutions are directed to consider energy efficiency when extending credit and loans to industry, making it less expensive for more efficient companies to get financing.[29] (This measure should be limited to financing for projects other than energy efficiency upgrades, lest it hamper the efficiency it is designed to reward.)

A government that prefers a more direct approach may reimburse companies for a share of the cost of energy efficiency upgrades, such as new boilers or waste heat recovery systems. For example, during the twelfth Five-Year Plan period, China offered RMB 200–300 (roughly $30–$43) in rebates for equipment per ton of coal equivalent saved by that equipment.[30]

One method of structuring financial incentives is to use a "cascade" approach, which relies on public money only to the extent necessary given market and project conditions:[31]

- For projects where commercial financing is available at reasonable rates and with reasonable oversight by lenders, government financing may be unnecessary.
- When commercial financing is not cost-effective, government may try to reform conditions in the financial system to remove structural barriers to financing and enable commercial lenders to take on these projects cost-effectively.
- For projects that are still too risky to be financed commercially at rates that are affordable for industry, government may partner with private

lenders using a risk-sharing instrument, which leverages government dollars to obtain some private sector funding.
- For projects of public value where it is impossible to interest commercial lenders even in a risk-sharing arrangement, government may consider being the sole provider of financial incentives.

Financial incentive programs should be coupled with robust monitoring and reporting of energy use data, to ensure that incentives are provided only when energy savings are realized.

Mandatory Targets

Governments may impose specific energy or carbon intensity targets for specific industries or for the economy as a whole. For example, in China's twelfth Five-Year Plan, the government required national energy intensity (energy per unit GDP) to decline 16 percent from 2010 to 2015 and industrial energy use per unit of value-added output to decline 21 percent over that period.[32] The central government identified more than 50 specific measures to be undertaken to accomplish these goals, many of which are implemented by provincial governments.[33]

Targets should be ambitious but technically achievable. One approach is to set targets by benchmarking against the most efficient facilities in the country or globally, ensuring that the technology to meet the targets exists in the commercial market. For example, in 2010 Japan introduced a program requiring industries to achieve a 1 percent annual energy efficiency improvement. Industries in major subsectors (e.g., steel, cement) were required to achieve the efficiency level of the top 10 percent or 20 percent of best-performing companies.[34] Similarly, the Netherlands set mandatory energy efficiency targets through negotiation with industry in its Energy Efficiency Benchmarking Covenant, aiming to fall within the top 10 percent of industries globally.[35] Unfortunately, the Dutch program did not reach its goal, as industrial energy efficiency improved at 0.5 percent per year, not enough to put Dutch industry into the top 10 percent.[36]

Targets should be technology neutral, allowing companies to determine the lowest-cost ways to reduce their energy intensities. Targets that reflect whole-facility performance may be particularly useful for rewarding companies that use integrative design, a set of principles that focus on achieving efficiencies through better linking of different components (e.g., via ductwork or piping).[37]

Opportunities for energy savings may be missed if policies consider only the efficiencies of the underlying components themselves (such as boilers) and not the energy losses as materials are moved between components.

Many governments have opted for voluntary rather than mandatory targets. These policies have similarities to mandatory targets, but they rely on providing participating companies with good publicity or other rewards to encourage compliance, such as eligibility for special tax breaks (an approach used in Denmark and the Netherlands). Voluntary targets are politically easier to enact but tend to be less effective than mandatory targets. An example of a voluntary program covering industrial efficiency (and methane leakage) is the United States' Natural Gas Star program.[38]

Encouraging the Adoption of Energy Management Systems

Governments may integrate energy management system requirements into national or subnational programs or codes. The existence of an international standard (ISO 50001) may simplify this process for many governments by reducing the need to devise and write specific energy management procedures or practices into law and by harmonizing requirements across jurisdictions (making it easier for multinational companies to comply). For example, Japan's energy conservation law makes specific reference to ISO 50001 as a compliance mechanism, and Canada offers a variety of incentives for ISO 50001–compliant businesses, such as cost-sharing assistance and training opportunities.[39]

Energy Audit and Energy Manager Requirements

Governments may require companies to undertake an energy audit, a process in which an energy management professional reviews current energy use and identifies opportunities for improvement. These requirements may be particularly useful for companies that want to participate in a government program that provides financing for upgrades, first to establish baseline energy use and improvement opportunities and later to verify that targets were met. Government may also establish requirements for energy managers or auditors. Several certification programs exist for energy management professionals, run by organizations such as the Association of Energy Engineers.

Policies to Support Research and Development

A variety of policies support businesses' efforts to improve their products and processes through R&D. A separate chapter in this policy design guide is devoted to R&D support policies, so they will not be discussed here.

When to Apply This Policy

Most countries, states, and provinces possess industrial facilities, some of which could probably benefit from industrial efficiency policies. However, the appropriate policies to encourage industrial energy efficiency depend on the level of industrialization and development of the region.

Lowest-Income Countries

Countries that have experienced little development may possess industries that rely significantly on human labor, such as natural resource extraction, cement and other construction materials, textiles and garments, and pulp and paper. Facilities are likely to have older and cheaper equipment, making them inefficient. This gives industries a large incentive to upgrade. However, capital and financing may be scarce, and more efficient equipment may have to be imported at high cost, making it difficult to afford upgrades. Government capacity to require accurate energy usage reporting or to enforce standards may be low.

In these regions, a combination of education or technical assistance and financing policies may be the best fit. Education about proper equipment maintenance and improved operations may deliver some energy savings even without capital investment. Many industries cannot afford to purchase upgraded equipment without aid and may have difficulty securing credit, so a government program to provide financing for efficiency upgrades may be necessary. A financing program should be coupled with technical assistance, provided by either the government or carefully vetted ESCOs, to ensure money is spent appropriately and the efficiency upgrades make sense. Some countries may be able to obtain international aid or development money to fund industrial energy efficiency programs.

Equipment standards may be considered for specific components, such as electric motors, which can be enforced at the point where the components are imported or manufactured. It may be advantageous to adopt a set of standards established by a more developed country with an expert regulatory agency rather than attempting to create voluminous technical standards from scratch.

Fiscal incentives and mandatory facility-wide or industry-wide targets may be poor choices. Fiscal incentives may be difficult to fund, and it may be challenging to ensure incentives are awarded only in proper circumstances. Facility-wide or industry-wide targets require monitoring capabilities that may be infeasible, and business owners might find that the necessary upgrades to meet the mandatory targets are financially out of reach without financing support.

Middle-Income Countries

This category includes developing countries that have made significant progress in diversifying their economies away from agriculture and natural resource extraction, with manufacturing and heavy industry often playing a large role. These countries are more urban and have higher energy consumption per capita than less-developed countries. It is particularly important that middle-income countries adopt strong industrial energy efficiency policies, because they are likely to have extensive industrial capital, much of it still inefficient.

Financial support and technical assistance remain useful in middle-income countries, although they are not as crucial as in less-developed countries. This is because more capital may be available through nongovernment channels, and more expertise may be available in the private sector. The inefficiency of much existing equipment already provides a significant financial incentive to upgrade.

Middle-income countries are at a stage where it is important to establish good practices in industry. Incentives or requirements for energy management systems, such as ISO 50001, may be introduced into national or subnational targets or policies. Similarly, requirements for energy audits and energy manager certification may ensure that energy savings are identified and realized efficiently.

Governments at this stage should be able to build functional energy reporting and monitoring programs, at least for larger industrial facilities. This opens up the possibility of using mandatory targets. Mandatory targets may be a good fit for a middle-income country, because their clarity and avoidance of technology prescription make them easier to comply with and to enforce. Financial incentives such as higher electricity prices for less efficient facilities may also be an option, as long as the energy monitoring system is resistant to cheating. Financial rewards for efficient equipment (such as high-efficiency boilers or air compressors) may also be a good fit if funding is available.

High-Income Countries

As a country develops, its economy may begin to transition from manufacturing to services. However, significant manufacturing is likely to remain, particularly for high-value goods and those that benefit from capital-intensive, highly refined production processes (e.g., those that involve robots and factory automation). The manufacturing of cars is a good example.

Financing and technical assistance programs are comparatively less important in advanced countries, because these countries have strong banking and financial systems that can provide credit to businesses seeking to upgrade their facilities. Similarly, they often possess high-quality training programs and the ability to attract skilled talent from other countries, reducing (but not eliminating) industry's need for education and technical assistance.

Equipment standards are highly effective in developed countries. These countries have the capability to develop robust, ambitious, and achievable standards and to gradually tighten those standards as technology develops. Mandatory targets at the facility level may also be helpful to encourage efficiency opportunities not captured by standards specific to individual pieces of equipment, such as those available through integrative design. Finally, developed countries are often in the best position to fund financial incentives, such as rebates for highly efficient equipment. Rebates and standards complement each other: Rebates provide an incentive to buy the most efficient models on the market, whereas efficiency standards tend to improve the efficiency of the bottom of the market by eliminating the worst performers.

Highly developed countries often import many of their manufactured goods. Because greenhouse gas emissions have global impact, an importer may be offshoring some of the emissions that are attributable to its consumption. Providing financial and technical assistance to developing countries that manufacture goods, thus facilitating their implementation of industrial energy efficiency programs, may in some cases allow emission reductions at lower cost than policies to improve the efficiency of domestic industry.

Detailed Policy Recommendations

The following policy design principles apply to industrial energy efficiency policies.

Policy Design Principles

Create a Long-Term Goal and Provide Business Certainty

It may take years for manufacturers of industrial equipment to research design improvements that reduce electricity or fuel consumption without adding too greatly to the equipment's cost and without compromising other performance characteristics. Businesses need sufficient time to make the necessary research and development investments and changes in their manufacturing lines. If standards are set just a few years at a time, businesses may be unsure whether standards will continue to become more stringent in future years. This means they will not know whether modest investment in near-term improvements is sufficient or if a larger investment in a new technology or major design change might be worthwhile.

Companies must often invest large amounts to build and equip facilities, which they then recoup by producing and selling products over many years. Long-term knowledge of the standards can help businesses sync their capital upgrade cycles to changes in the standards.

Financial incentives and especially financing programs also need to have a long time horizon so that industries can take advantage of these programs when making capital equipment budgeting plans (which may extend years into the future). Certainty of the availability of financing lowers risk, aiding business owners' calculations about what upgrades make sense and potentially saving money or improving shareholder confidence.

Build In Continuous Improvement

Continuous improvement is especially important for efficiency standards. The purpose of efficiency standards is to eliminate the worst-performing products from the market, gradually improving the efficiency of industry as a whole. If standards are not updated to become more stringent over time, they become ineffective at driving improvements and fail to achieve their purpose. Standards may be reviewed at known intervals, or improvements may be based on the top-performing industrial equipment already commercially available.

Focus Standards on Outcomes, Not Technologies

To the extent feasible, given the diversity of types of industrial equipment, standards should be written to be technology neutral. For example, standards should be based on performance characteristics of the type of equipment being regulated rather than design characteristics of the equipment. This policy design principle also applies to mandatory targets, which should set overall energy use requirements for facilities or per unit output, to give businesses flexibility in identifying technologies and processes to be used to achieve the target.

Ensure Economic Incentives Are Liquid

Businesses sometimes fail to achieve profits. This may be due to a period of low sales, a decision by management to reinvest revenues in expanding manufacturing capacity, or other factors. If financial incentives for efficiency upgrades are provided in the form of nonrefundable tax credits, businesses may be unable to take advantage of the incentives if they do not have enough taxable income after deductions. Incentives provided as a rebate, cash grant, or low-interest loan are accessible to a broader range of manufacturers.

Additional Design Considerations

Avoid Inducing Leakage and Offshoring

The industry sector is more able than other sectors (transportation, building, electricity generation, or land use) to respond to tougher regulations by moving operations to a different jurisdiction to escape the reach of the policy. This is commonly called leakage or offshoring. Policymakers designing policies to encourage industrial energy efficiency must therefore be mindful about imposing financial burdens. For policies that provide "carrots" such as low-cost financing, technical assistance, rebates on efficient equipment, and the like, inducing leakage is not a concern. Well-designed equipment standards will mandate the purchase of equipment with reasonable payback periods, a beneficial investment that generally will not encourage offshoring. Financial penalties for industries that use a lot of energy are the type of policy most likely to cause leakage.

The decision to relocate production is a major one, and many factors play into a company's decision (e.g., the availability of skilled labor, inputs, shipping costs). Policies to encourage industrial efficiency, even those that impose modest penalties on inefficient industry, are unlikely to be a primary cause of relocation. Engaging with industry stakeholders while designing policy and introducing "carrots" and "sticks" as a single package may help to gain industry buy-in and reduce leakage risk.

Over the long term, the best method to reduce offshoring is to develop a local environment that is favorable to industry. For example, robust investment in infrastructure and training programs to ensure a ready supply of skilled workers (such as Germany's Dual Vocational Training Program)[40] can help ensure that industries are better off if they stay and invest in efficiency than if they move to other regions where they need not invest in efficiency.

Case Studies

China's Top-10,000 Program

The largest energy-saving policy program in the world is the "Top-10,000 Program," an effort launched by China as part of its twelfth Five-Year Plan to reduce industrial energy consumption. The program targets "16,078 enterprises with annual energy consumption of approximately 2 billion tons of coal equivalent, accounting for roughly 87 percent of China's total industrial energy consumption and 60 percent of the total national energy consumption."[41]

The program incorporates many specific policies. Examples include training programs for energy managers, energy audits and an energy use reporting system, specific energy use reduction targets by province, both financial incentives and financing support for energy efficiency retrofits, and promoting work with ESCOs. Some enterprises also have energy-saving targets and are subject to sanctions if the targets are missed.[42] The program's goal was to save at least 255 million tons of coal equivalent during the twelfth Five-Year Plan period of 2011–2015.[43] From 2010 to 2014, China's industrial energy use rose by 28 percent, but this was slower than the rate of economic growth. China's energy use per unit GDP dropped by 14 percent over the same period.[44] As a result, China's National Development and Reform Commission announced that the Top-10,000 Program saved 309 million tons of coal equivalent during this period, or 121 percent of the program's target.[45]

A program that can achieve reductions in total industrial energy use, not just energy intensity of the economy, will be important for China in the years to come.

United States's Superior Energy Performance Program

The Superior Energy Performance program is a voluntary program in the United States to provide assistance and give recognition to industries that improve their fuel efficiency. The program began in a pilot phase in 2010 and was launched nationally in 2012. To participate, a company must implement an energy management system and comply with a standard set by the International Organization for Standardization. The company must demonstrate at least 5 percent energy savings (for companies new to energy management) or 15 percent over 10 years (for companies with a longer track record of energy management). The level of achieved improvement in energy efficiency determines whether a company is certified at the "Silver," "Gold," or "Platinum" tier.[46]

The program includes a strong education and training element. In-person training and web-based tools and resources are available through the program.[47]

Facilities have achieved 5.6 to 30.6 percent improvement in energy efficiency over 3 years. The payback period is typically under 1.5 years for facilities with energy costs of at least $2 million per year.[48]

The Superior Energy Performance program is now being complemented by a new "50001 Ready" Program.[49] This program has lower requirements and aims to entice more participants than the original program.

The main weakness of the Superior Energy Performance program is its voluntary nature. In fact, except for equipment efficiency standards, all U.S. programs for industrial energy efficiency are voluntary. (Only new source performance standards, under the Clean Air Act, require compliance.[50] However, for industrial facilities, these standards pertain to emissions of non-CO_2 pollutants,[51] and there are many means by which these emissions can be reduced other than by improving energy efficiency.)

Bulgarian Energy Efficiency and Renewable Sources Fund

In 2004, Bulgaria established a revolving fund for financing energy efficiency projects. Originally called the Bulgarian Energy Efficiency Fund, it was renamed the Bulgarian Energy Efficiency and Renewable Sources Fund (EERSF) in 2011, when distributed renewable energy projects became eligible for funding. Initial capital for the fund came from the World Bank's International Bank for Reconstruction and Development, the governments of Austria and Bulgaria, and private companies in Bulgaria.[52] The fund assists industry via three mechanisms:

- The fund acts as a lender, directly providing financing to industries. Typical interest rates are 4.5 to 9 percent, and financed projects must use well-proven technologies with payback periods of 5 years or less.[53]
- The fund can provide credit guarantees, reducing the risk to private lender and thereby helping companies get financing from private banks. Credit guarantees cover either 50 or 80 percent of the total value.[54]
- The fund may act as a consulting company, directly providing technical assistance to Bulgarian firms on energy efficiency projects.[55] EERSF may do this through collaboration agreements with ESCOs. It has signed such agreements with 17 ESCOs.[56]

Through December 2014, the fund "funded or provided guarantees to 170 energy efficiency projects for a total amount of 45.8 BGN (23.4 M €)." A total of 160 of those projects, which were funded by December 2013, achieved 95.4 gigawatt-hours per year of energy savings.[57]

Conclusion

The industrial sector is responsible for the plurality of the world's energy consumption and greenhouse gas emissions. Reducing energy use in industry is a crucial part of the transition to a clean energy economy, and policymakers can take steps to accelerate this process. Five key types of policies, used in combination, deliver the best results: education and technical assistance, financing, financial incentives, mandatory targets, and equipment standards. Fundamentally, these policies help industry invest in higher-quality equipment that will save energy and quickly pay for itself, increasing the long-term competitiveness of affected businesses. As policies that both strengthen the economy and lower emissions, industrial efficiency policies are well suited to nearly all countries, especially those in need of ways to cut emissions while promoting long-term economic development.

CHAPTER TWELVE

Industrial Process Emission Policies

Industrial process emissions reflect all the nonenergy ways in which industrial production results in the release of greenhouse gases into the atmosphere. For example, natural gas leaks from pipelines and methane produced by enteric fermentation in livestock are types of process emissions. For purposes of this chapter, agriculture and waste management are each treated as an industry, although some reports classify these economic activities differently. Whereas the majority of energy-related emissions consist of carbon dioxide (CO_2), process emissions are a mixture of CO_2, methane (CH_4), nitrous oxide (N_2O), and various fluorinated gases (F-gases) with high global warming potentials, often used as refrigerants and propellants.

In many countries, process emissions account for a large share of all industrial greenhouse gas emissions. Figure 12-1 shows industry process emissions in the context of direct energy-related industry emissions, industry's share of emissions from the production of electricity and heat, and nonindustry emissions for the United States and China.

In the United States in 2016, process emissions were 1.6 billion metric tons carbon dioxide equivalent (CO_2e), representing 55 percent of all industry-related emissions and 24 percent of economy-wide emissions. In China, process emissions were 2.6 billion tons, although this represented only 28 percent of industry emissions and 20 percent of economy-wide emissions. The share relative to total economy-wide emissions is smaller because industries in China are less energy efficient than those in the United States, China has more heavy industry in its industrial mix, and a larger share of China's final energy use is generated by coal.

Part of the challenge in reducing process emissions is the diversity of ways in which the emissions are generated. Although there are many cross-cutting techniques that save energy in a variety of different industries, measures to reduce process emissions often must be specific to each type of process.

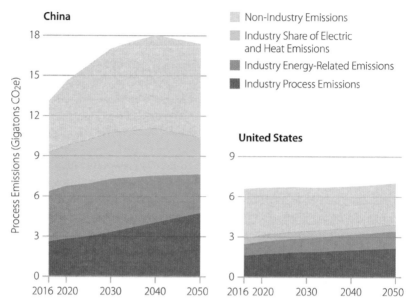

Figure 12-1. U.S. and China business-as-usual process emissions in context. (From Energy Policy Simulator, Energy Innovation, January 10, 2018, https://energypolicy.solutions.)

Fortunately, a limited number of processes are responsible for most process emissions. For example, Figure 12-2 shows that nine sources of process emissions were responsible for nearly 90 percent of all process emissions globally in 2010. Given the high concentration in a small set of processes, a limited number of technologies and strategies can have an outsize impact on process emissions.

Tackling process emissions can achieve at least 10 percent of the greenhouse gas emission reductions necessary to hit the two-degree target (Figure 12-3).

Policy Description and Goal

Given the fact that most of the world's process emissions are produced from a narrow range of activities, our efforts will be most efficient and will have a greater impact if we focus on specific types of emissions: enteric fermentation and manure from livestock; methane leaks from natural gas systems; cement clinker production; greenhouse gases from soils, rice cultivation, and fertilizer use; methane produced in landfills; refrigerant use; methane leaks from

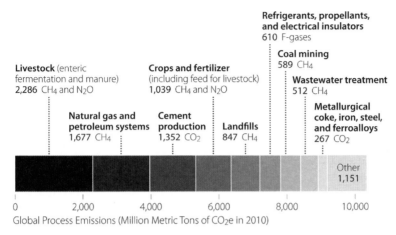

Figure 12-2. Nine processes dominate global process emissions. ("Global Anthropogenic Non-CO$_2$ Greenhouse Gas Emissions: 1990–2030," U.S. EPA, 2012, https://www.epa.gov/global-mitigation-non-co2-greenhouse-gases/global-anthropogenic-non-co2-greenhouse-gas-emissions; "Climate Change 2014: Mitigation of Climate Change," 2014, 749.)

coal mines; methane from wastewater treatment; and metallurgical coke for iron and steel production. Techniques exist to tackle each of these types of emissions, and policies such as monitoring and reporting requirements, performance standards, carbon pricing, and financial and technical assistance can help achieve emission reductions from each of these processes. Examples of successful programs include the Montreal Protocol (which in 2016 was expanded to cover refrigerants) and methane reduction initiatives to promote the use of anaerobic digesters (discussed later in this chapter) in California and Zimbabwe.

Technologies and Measures to Reduce Process Emissions

This section discusses technologies and measures to tackle process emissions from each of the sources depicted in Figure 12-2, in order from greatest to least global process emissions.

Livestock Measures

Animal husbandry produces greenhouse gas emissions through two mechanisms: the decomposition of manure and enteric fermentation. If manure is left unmanaged, bacteria that consume nutrients in the manure release methane as a byproduct of their metabolism. The primary method of reducing these

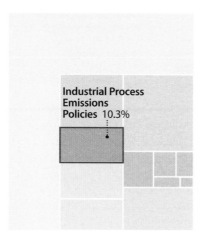

Figure 12-3. Potential emission reductions from industrial process emission policies. (Analysis done using data with permission from the International Institute for Applied Systems Analysis [IIASA]. Data source: Tavoni et al., 2013. Data downloaded from the LIMITS Scenario database hosted at IIASA, https://tntcat.iiasa .ac.at/LIMITSPUBLICDB/dsd?Action=htmlpage &page=about.)

emissions is to process manure in anaerobic digesters, which convert the produced methane into electricity. This benefits the environment and provides power for farm or dairy operations. Anaerobic digesters are a commercialized technology, making them a promising method for achieving emission reductions from livestock. For example, in 2016 California passed legislation that will regulate greenhouse gas emissions from livestock beginning in 2024, and regulators are focusing on deploying anaerobic digesters as a means to hit the state's target.[1]

The other major source of livestock methane is enteric fermentation. Certain herbivores, called ruminants, have two stomachs. Plant matter chewed and swallowed by a ruminant is sent to the first stomach, where bacteria help to break down the food in a process called fermentation. Methane is released as a byproduct, which is then expelled by the animal, mostly by belching. The animal later moves the plant matter back to the mouth, rechews it, and sends it to the second stomach, which leads onward to the rest of the digestive system. Ruminants convert more of the energy in their feed into methane than nonruminant animals. For example, a ruminant such as a cow may convert 5.5–6.5 percent of the feed energy into methane, whereas that percentage is closer to 2.5 percent for horses and 0.6 percent for swine (both nonruminants).[2] Because of the large number of cattle in agriculture systems worldwide, the large amount of feed consumed by each cow, and the high methane conversion percentage for ruminants, cattle are responsible for most methane produced by enteric fermentation.

Various techniques have been proposed for reducing methane from enteric fermentation, such as varying the type, schedule, or quantity of food given to animals; including supplements in the food; or even vaccinating the animals so their immune systems attack methane-generating bacteria. None of these practices are yet well established or used commercially. Finally, a more direct way to reduce emissions from both manure and enteric fermentation would be to reduce the number of animals (particularly ruminants) in the agriculture system by reducing demand for meat and dairy products.

Prevent Methane Leaks from Natural Gas and Petroleum Systems

Methane, a powerful greenhouse gas, is the primary component of natural gas. It is colorless and scentless. (The familiar smell of natural gas is an odorant that is added for safety reasons.) Natural gas can leak at any point in its economic lifecycle: from the wellhead where it is produced, pipelines and meters that carry the gas to customers, customer-owned pipes and valves behind the meter, or gas-fired appliances.

Natural gas leaks are difficult to eliminate. A single oil and natural gas field may have up to a million connections (e.g., joints between pipes, gaskets, valves), and if even a small percentage of them are not perfectly airtight, high-pressure gas can be forced out.[3] Natural gas pipelines in cities can be decades old and buried underground, making detection and repair of leaks difficult. For example, the City of Boston loses $1 billion worth of natural gas each decade though leaks from its aging cast-iron pipelines.[4]

Nonetheless, a variety of techniques can be used to detect and minimize methane leakage. Infrared video cameras can visualize methane plumes, and oil field operators may regularly survey their equipment with camera-equipped vehicles to check for leaks. Technology exists to capture the gas in the flowback from a newly drilled well, a process called green well completions.[5] (More information on this appears in the "Case Studies" section later in this chapter.) And in the United States, studies have found that a disproportionate amount of leaked methane from oil and natural gas production facilities comes from a small number of problem sites,[6] providing a chance to reap outsized benefits from tackling these opportunities. In some cases, fixing leaks saves money, because the gas being lost to the atmosphere is the product being transported and sold to end users.

Cement Clinker Substitution

Cement is an essential component of concrete, one of the most important building materials used around the world. Clinker is a hard substance that composes 74–84 percent of cement by weight, varying by world region.[7]

Clinker is created by the breakdown of calcium carbonate ($CaCO_3$), the main constituent of limestone rock, into lime (CaO) and carbon dioxide (CO_2). The CO_2 is released to the atmosphere, and the lime reacts with silica, aluminum, and other materials to become part of the clinker.[8]

The main method by which cement clinker emissions can be mitigated is to reduce the share of clinker in cement. Other materials such as fly ash (a waste product from coal combustion) may be substituted for a portion of the clinker. However, there is a limit to how low the clinker percentage may be before the cement has undesirable structural properties. It is estimated that the clinker percentage may be reduced to as little as 70 percent.[9]

Another way to reduce emissions is by improving building quality and longevity, particularly in developing countries where construction standards may be lower than in developed countries. For example, in China, "the average period of time that . . . housing remains livable is a meager 35 years compared to a century or more in many developed countries," and many buildings last no more than 20 years.[10] If buildings, roads, and other infrastructure need to be rebuilt every few decades, this requires much more cement production and hence results in more process emissions than if buildings last a century.

Cropland and Fertilizer Management

Growing crops generates greenhouse gas emissions through the decomposition of organic matter in the soil and the application of fertilizer.

Each year, organic matter is added to the soil as crops grow, and it later decays, releasing CO_2 and methane. Tilling (turning over and breaking up the soil) is often performed before planting, to loosen compacted soil and mix nutrients. However, this process exposes organic matter to air and accelerates the release of emissions into the atmosphere. Tilling the soil less frequently or refraining from tilling altogether can allow more carbon to build up in the soil until the soil becomes saturated with carbon in about 20–25 years. However, reducing or eliminating tilling can reduce the productivity of some crops.[11]

After the harvest each year, planting grass or a legume as a cover crop for the winter can help reduce the release of organic matter from soils. Also, a winter cover crop may allow less fertilizer to be used in the next season.

Reduced tillage and the use of winter cover crops must be continued for as long as the carbon storage is to be maintained. Should the cropland be returned to typical farming practices, the stored carbon will be released to the atmosphere.

Application of fertilizer results in emissions of N_2O, because only some of the nitrogen in the fertilizer is successfully taken up by the plants. There are techniques to increase the fraction of applied nitrogen that is used by plants, including improved fertilizer application methods and timing, reduction in the amount of fertilizer used, and the use of nitrification inhibitors. Nitrification inhibitors are chemicals that slow the conversion of ammonium (NH_4+ in decomposed fertilizer) to nitrate (NO_3-, a plant-available form of nitrogen), so that plants have time to capture more of the nitrate before it volatilizes or leaches away.

Additionally, there are several considerations particular to the growing of rice. Around the world, most rice is grown in paddy fields, which are flooded with water periodically throughout the growing season. While a rice field is flooded, the soil is in an anaerobic environment, or deprived of oxygen. This allows microbes to produce methane as they ferment organic matter in the soil.

A key method of reducing methane emissions from rice cultivation is to reduce the amount of time the rice fields remain flooded. Draining a field halfway through the growing season or engaging in alternating wetting and drying greatly reduces methane emissions, and these practices may or may not decrease crop yields, depending on the soil conditions, climate conditions, wetting-and-drying technique, and rice cultivar.[12] However, these water management practices may not be possible in some areas that are naturally flooded, where farmers lack reliable control over irrigation systems, or where fields are not sufficiently level (because of the formation of wet and dry pockets).

Preventing Methane Leaks from Landfills

The decomposition of organic material under anaerobic conditions in landfills results in the generation of methane, which makes up roughly 50 percent of landfill gas.[13] Rather than being allowed to reach the atmosphere, landfill gas can be harvested by drilling wells in the landfill and using a blower or vacuum system. This allows the gas to be collected at a central point, where it can be used to generate electricity (usually by powering an internal combustion engine) or may be used directly to replace another fuel, such as natural gas or coal.[14] Another way to reduce landfill methane emissions is to divert organic waste from landfills through techniques such as reducing food waste and composting.[15] (During composting, decomposition typically occurs in aerobic conditions, which produces much less methane than anaerobic decomposition.)[16]

Eliminating Refrigerants with High Global Warming Potential

Fluorinated gases (F-gases) are chemicals used for a variety of industrial purposes. For example, F-gases are often used as a refrigerant, as a propellant in aerosol canisters, and as an electrical insulator in high-voltage transmission systems. Many F-gases are replacements for ozone-depleting substances, chemicals that damaged Earth's ozone layer and that were largely phased out as a result of the 1987 Montreal Protocol. Although the remaining F-gases used in industry today do not damage the ozone layer, many are powerful and long-lived greenhouse gases, so they contribute significantly to global warming.

Most F-gases that are produced eventually reach the atmosphere. For example, the refrigerant in an air conditioner may leak slowly, or it may be released when the air conditioner is scrapped at the end of its useful life. One way to reduce F-gas emissions is to better seal systems against leaks and to establish recycling or takeback programs for old products containing F-gases. However, the best technique is to avoid creating F-gases in the first place by substituting different chemicals in industrial applications. The best choices are chemicals that neither contribute to global warming nor damage the ozone layer but still allow the system to operate at high efficiency. Examples of environmentally friendly refrigerants include R-717 (ammonia), R-744 (CO_2), R-1270 (propylene), R-290 (propane), R-600a (isobutane), and R-1150 (ethylene).[17] Some of these alternatives come with their own dangers (e.g., propane is highly flammable), but different substitutes can be used for different applications to lower risks.

The 2016 Kigali amendment to the Montreal Protocol, discussed as one of the case studies in this chapter, mandates the phase-out of many F-gases and the substitution of less harmful refrigerants and propellants.

Controlling Methane Leaks from Coal Mining

As coal is formed underground over millions of years, methane is also formed in the coal seams. Coalbed methane refers to all of the methane that forms in these seams, whereas coal mine methane refers to the portion of that methane that would be released through mining activities.[18] In underground mines, coal mine methane buildup can create an explosive hazard and a danger to miners, so it is collected and vented with degasification systems. Methane also can leak from mine ventilation systems, and leaks may persist after a coal mine is abandoned.[19]

Equipment may be used to capture the methane emitted from coal mines, particularly from degasification systems, which emit methane at a higher concentration than ventilation systems. Mine ventilation systems are the single largest source of methane emissions from coal mines, but the high air flow rate and low methane concentrations (less than 1 percent) make it difficult to capture and use this methane cost-effectively.[20]

Captured coal mine methane can be put to economically productive uses. Captured coal mine methane is most often used for power generation, district heating, or boiler fuel, or it can be used on-site for coal drying, fueling mine boilers, or other purposes.[21]

Finally, technologies that reduce the use of coal as an energy source (such as renewable energy sources) might allow for fewer coal mines and thereby reduce coal mine methane emissions.

Reducing Methane Emissions from Wastewater Treatment

Methane is produced from decomposing organic material, especially under anaerobic conditions. In developed countries, most wastewater is treated in aerobic conditions, so little methane is generated directly. However, the biosolids that remain after water treatment, if not managed properly, may produce methane. In developing countries, wastewater, if treated at all, is usually treated under anaerobic conditions and produces methane directly.[22]

Biosolids may be processed in an anaerobic digester. (Anaerobic digesters are discussed earlier, in the "Livestock Measures" section, and later as a case study.) In developing countries without modern wastewater treatment facilities, the best solution is to construct centralized aerobic water treatment plants, if the population and infrastructure can support these facilities. Otherwise, existing anaerobic lagoons may be retrofitted with covers and biogas capture systems, a simple and low-cost measure. Additionally, it is important to ensure staff are trained to maintain and efficiently operate the facilities.[23]

The captured biogas may be treated and sold to a natural gas utility or used as fuel for fleet vehicles. Captured biogas can also be burned to generate electricity or heat, ideally in a combined heat and power system.[24]

Reducing Metallurgical Coke Production for Iron and Steel

Most iron and steel is created in blast furnaces or basic oxygen furnaces. These furnaces are used to convert mined iron ore into pig iron or to convert pig iron and various alloy metals into steel. Blast furnaces are fueled by coke, a substance created when pulverized coal is heated in an oxygen-free environ-ment, a process called coking.[25] Coke is used both as a chemical-reducing agent (to remove oxygen from the iron oxide in iron ore) and to generate the high temperatures needed for steelmaking.[26]

Several techniques exist for reducing coke production and use. Electric arc furnaces can generate the high temperatures needed to make steel without the need for coal, but they do not provide a source of carbon as a chemical-reducing agent, so they are used primarily to reuse scrap steel rather than to create new steel.[27] Modern blast furnace designs can reduce the amount of coke needed to produce a quantity of steel. And steel companies are develop-ing technology that would allow natural gas to be used in place of coke as the carbon source in steelmaking, thereby offsetting the emissions generated when producing coke.[28]

Policies to Reduce Process Emissions

A variety of policies can be used to tackle process emissions. Many are dis-cussed in detail in other chapters of this book and therefore will be only briefly covered here.

Monitoring and reporting requirements may be a necessary precondition for the application of certain other policies, such as performance standards, financial incentives, or a carbon price. Currently, few industrial facilities mea-sure or track their process emissions. Where direct measurement may not be practical (e.g., in agricultural operations), equations may be developed that estimate process emissions based on the amount and type of activity taking place (e.g., the number of cattle in a dairy, what they are fed, how their manure is managed).

Performance standards may be most familiar in the context of regulating energy use (such as the distance a car can travel on a given quantity of gaso-line), but they can also be applied to process emission sources. For example, government may mandate that a certain percentage of landfill gas be captured from landfills or that no more than a certain percentage of clinker be used in cement. In the natural gas and petroleum industry, standards may mandate green well completions, which reduce methane leakage at the wellhead. Coal mines may be subject to requirements for capture and flaring of methane. To control F-gases, governments may ban specific refrigerants and propellants. Standards may also be used to tackle building quality and longevity, lowering the amount of cement needed for reconstructing buildings.

A carbon price (a carbon tax or cap-and-trade program, discussed in Chapter 13) can provide a financial incentive for companies to find cost-effective ways to reduce process emissions. Of course, to be effective the carbon price must be applied not only to energy-related emissions but also to process emissions, and it must be levied not only on CO_2 but also on other greenhouse gases.

Financing assistance can help companies purchase different or upgraded equipment, such as an anaerobic digester for a dairy, or retool their assembly lines to use substitutes for F-gases in refrigerators and air conditioners. Financing could be provided via grants, low-interest loans, loan guarantees, a revolving fund, a green bank, or other mechanisms.

Government procurement decisions can account for lifecycle emissions, including process emissions. If government refuses to purchase goods whose manufacture entailed high emissions (or applies a "shadow price" to these goods, making them less able to compete with goods produced in a more environmentally friendly way), this encourages emission cuts by companies that want to sell their goods to the government. An example of this type of policy is California's Buy Clean Act.[29]

Technical assistance and training programs can help some businesses, particularly small and medium enterprises and farmers, to understand their sources of process emissions and mitigation strategies. Information about best practice inspection and maintenance protocols can help avoid degradation in performance over time. Technical assistance may be particularly helpful in developing countries and when dealing with industries containing many small producers, such as agriculture (livestock measures, cropland management), and in some countries, the cement and waste management industries.

Economic signals for goods or materials substitution may be used to divert purchases from high-emission to lower-emission choices. For example, a tax on virgin steel or a subsidy for recycled steel may increase the share of steel production from scrap steel in electric arc furnaces. A tax on meat from ruminant animals may encourage a shift to nonruminant animal or plant sources of protein. And a variety of policies (discussed elsewhere in this book) can encourage a shift from coal-fired power generation to other electricity sources.

When to Apply This Policy

Because almost every country has agricultural or industrial production, it is likely that almost every country could benefit from application of some of these policies. The choice of policies to focus on in a particular country should be guided by the prevalence of relevant industries and practices that could be changed.

For example, in countries with rapid urbanization and building quality problems, such as China, codes mandating building and infrastructure quality, to promote longevity, may be among the highest-priority policies. It is cost-effective to put strong building codes in place while a country is still expanding its infrastructure base and building stock, to avoid the cost and emissions associated with rebuilding. In a country that is already urbanized and has a high-quality building stock, these policies are less important.

Countries with large amounts of natural gas production and old gas distribution infrastructure, such as the United States, are prime targets for policies tackling methane leaks. Countries with large coal mining operations should consider ways to tackle coal mine methane. Any country that produces refrigerants and propellants should use policy to mandate a transition from F-gases to safer chemicals.

Case Studies

Policies to tackle process emissions are novel, but some helpful precedents do exist.

The Montreal Protocol

Background

The Montreal Protocol is a landmark international agreement finalized in 1987 and amended several times since then, most recently in Kigali in 2016. In this treaty, countries agreed to phase out ozone-depleting substances that were progressively destroying Earth's stratospheric ozone layer, which protects the surface from harmful radiation.[30] The treaty was successful at reducing emissions of more than 100 ozone-depleting substances by more than 99 percent.[31]

The protocol is widely regarded as the most successful international environmental treaty. In fact, United Nations secretary-general Kofi Annan stated that it was "perhaps the single most successful international agreement to date" (i.e., environmental or otherwise).[32] It was the first universally ratified treaty in United Nations history.[33]

Originally, the Montreal Protocol primarily targeted two categories of gases:

chlorofluorocarbons (CFCs) and hydrochlorofluorocarbons (HCFCs), both of which damage the ozone layer and are also potent greenhouse gases. CFCs, the more harmful of the two categories of gases, were to be phased out on a faster schedule than HCFCs. The protocol was amended several times to add new chemicals to the list of controlled substances and to accelerate the phase-out timetables for chemicals already on that list.[34]

In 2016, under the Kigali Amendment, the agreement was modified to require the phase-out of hydrofluorocarbons (HFCs).[35] This class of chemicals is not damaging to the ozone layer, but HFCs are potent greenhouse gases, and they are used for some of the same purposes (such as refrigerants and aerosol propellants) as were CFCs and HCFCs before them. This amendment, and actions taken by individual countries to comply with it, may be the best policy option for reducing F-gas process emissions globally.

Reasons for Success
There are several reasons for the Montreal Protocol's success. First, the science behind ozone layer depletion was widely accepted by the general public. After the 1973 discovery that CFCs could lead to ozone depletion, the industries that manufactured CFCs and aerosols, particularly DuPont, undertook a massive disinformation campaign in an attempt to cast doubt on the link between their products and ozone layer damage.[36] However, the discovery of the ozone hole above Antarctica in 1985[37] helped to confirm the science and demonstrate the urgency of the problem, catalyzing the international community.

Second, the Montreal Protocol included a multilateral fund, which provided a way for developed countries to provide financial support to help developing countries transition away from CFC and HCFC use. This fund is still important today in facilitating the phase-out of the last HCFCs still being produced in a few developing countries.[38]

Third, the treaty avoided becoming politicized. In the United States, it was Republican president Ronald Reagan's crowning environmental achievement, and it has been supported and strengthened by both Republican and Democratic presidents.[39]

Examples of Country-Level Implementation: The United States and Japan
As an international treaty, the Montreal Protocol had to be made effective through laws in each individual country. This is where the policy design guidelines in this chapter come in.

For example, in the United States, the Environmental Protection Agency (EPA) implemented the HFC phase-out through the Significant New Alternatives Policy program. Under this program, the EPA evaluated alternative chemicals for a variety of different use cases, such as chemicals used for fire suppression, household refrigerators, and motor vehicle air conditioning systems.[40] Factors considered by the EPA include effects on ozone depletion, climate change, exposure and toxicity to humans, flammability, and other environmental impacts. The EPA uses a comparative risk framework: It does not require that substances be risk-free, and it restricts only chemicals whose impacts are significantly worse than alternatives, to avoid interfering with the market's selection of alternative substances more than necessary.[41]

Japan has been a leader in using policy to accelerate the transition away from F-gases. Even before the Kigali Amendment, Japan enacted policies to begin drawing down refrigerant emissions.[42] Japan's law, enacted in 2015, covers the entire product lifecycle. It phases down refrigerant production by manufacturers and importers, promotes low–global warming potential and nonfluorocarbon products, prevents leakage from equipment while in use by requiring periodic checks and maintenance, and promotes recycling by licensed companies at end of equipment life.[43] Japan worked with 14 domestic industry organizations, such as the Japan Chemical Industry Association and the Japan Automobile Manufacturers Association, to implement product labeling with refrigerant and global warming potential information and to encourage manufacturers to set and achieve voluntary targets.[44]

Zimbabwe's National Domestic Biogas Program and California's Dairy Digester R&D Program

Of the process emission reduction technologies discussed in this chapter, anaerobic digesters for manure and organic waste are among the most broadly applicable. Government programs to fund digester construction exist in places ranging from California, one of the world's most developed and high-tech economies, to rural Zimbabwe. Both of these programs are discussed here to illustrate the range of approaches that can be used to successfully deploy this technology.

In California, the state's Department of Food and Agriculture operates the Dairy Digester Research & Development Program, which provides financial assistance for the installation of digesters in dairies in California.[45] The program is funded by the state's Greenhouse Gas Reduction Fund.[46] In 2017, 18 projects were awarded a total of $35.1 million in funding, with a total project cost of $106.4 million, so every government dollar leveraged more than $3 in private sector money.[47] Project sponsors must submit a detailed application and are carefully evaluated against a variety of criteria, including benefits to disadvantaged communities, financial strength, greenhouse gas emission reduction estimates, economic and environmental co-benefits, and project readiness (including compliance with the California Environmental Quality Act and other laws, as well as outreach to surrounding communities). Awardees must report verified emission reductions for 10 years after the project is operational.[48]

In Zimbabwe, three government ministries partnered with two Dutch non-government organizations, SNV and Hivos, to create the National Domestic Biogas Programme.[49] This program aims to construct digesters throughout Zimbabwe to facilitate the creation of biogas (a mixture of gases from decomposition of organic material, primarily methane). Rather than being used to generate electricity, as in California, the biogas created in Zimbabwe will be used directly for cooking and lighting and for powering biogas-burning appliances.[50] Anaerobic digesters have many benefits when used in rural Africa. In addition to reducing emissions, a digester may improve the sanitary conditions on a farm, eliminate indoor air pollution from the burning of traditional biomass, significantly reduce the workload related to food preparation, and produce high-quality fertilizer (the fermented bio-slurry).[51] However, anaerobic digesters need significant amounts of manure and water to operate effectively, so they may not be a fit for parts of Africa that lack access to reliable sources of usable water.[52]

To date under this program in Zimbabwe, more than 70 masons and 18 fabricators of parts have been trained in how to construct biogas plants.[53] The project has supplied more than 1,385 households with biogas,[54] has a target to construct 8,000 digesters by 2018,[55] and aims to establish a vibrant, local biogas sector that will ultimately serve more than 67,000 households.[56]

U.S. Regulations to Reduce Methane Leaks from the Natural Gas and Petroleum Industry

In 2012, the EPA enacted standards limiting the emissions of volatile organic compounds, a type of local air pollutant, from natural gas and petroleum operations. In 2016, this rule was extended to cover greenhouse gases, particularly methane, and to cover a greater range of industrial activities and equipment. The new rule also requires owners and operators to find and repair methane leaks.[57] The rules require monitoring for leaks on a fixed schedule (twice a year or quarterly, depending on the type of equipment). The rules allow multiple leak detection methods, including optical imaging or the use of an organic vapor analyzer,[58] and they permit producers to apply to use other technologies, to allow innovation.[59]

The rule also phases in requirements that drillers perform green well completions. After a new well is drilled, before it begins production, a substance called flowback (a mixture of drilling fluids, gas, oil, water, and mud) is extracted from the well. This process takes from a single day to several weeks. (Wells involving hydraulic fracturing, or "fracking," produce more flowback than non-fracked wells.) In a conventional well completion, the flowback is directed to an open pit or tank, where the gas released from the mixture is vented to the atmosphere, or sometimes flared. In a "green completion," equipment is used to separate the gas, liquid, and solid components of the flowback mixture, and the gas is captured for onsite use or sale.[60] The necessary technology is mature and can capture up to 90 percent of the gas in the flowback stream.[61]

The EPA has estimated that the rule will yield climate benefits of $690 million in 2025, outweighing the costs of $530 million, and will also achieve improvements in public health (reduced illness and deaths) thanks to reductions in toxic air pollutants.[62]

Conclusion

Industrial process emissions are an important source of greenhouse gases, accounting for 55 percent of all industry-related emissions in the United States and 28 percent in China. Although industry is diverse, just nine types of activity produce almost 90 percent of global process emissions: enteric fermentation and manure from livestock; methane leaks from natural gas systems; cement clinker production; emissions from soils, rice cultivation, and fertilizer use; methane produced in landfills; refrigerant use; methane leaks from coal mines; methane from wastewater treatment; and metallurgical coke for iron and steel production. Techniques exist to tackle each of these types of emissions, and policies such as monitoring and reporting requirements, performance standards, carbon pricing, and financial and technical assistance can help achieve emission reductions. Examples of successful programs include the Montreal Protocol (which in 2016 was expanded to cover F-gases) and anaerobic digester deployment initiatives in California and Zimbabwe.

CROSS-SECTOR POLICIES

In addition to sector-specific policies, cross-sector policies are crucial to decarbonizing the economy. Indeed, one of the most important policies for decarbonization, carbon pricing, typically operates across multiple sectors, helping deliver large emission reductions. Similarly, support for research and development (R&D), which is critical to lowering the long-run costs of decarbonization, typically targets technological breakthroughs in different parts of the economy.

These policies are essential for decarbonizing the economy and doing so cost-effectively. Although the effect of carbon pricing is directly related to the price or emission cap used, in our analysis we find that strong carbon pricing set at the social cost of carbon can achieve 26 percent of the emission reductions necessary by 2050 to hit the two-degree target.

We do not explicitly model the effect of R&D on reducing emissions because of the challenges in making assumptions about R&D achievement and spending. However, R&D breakthroughs would lower the costs of meeting the two-degree target and would probably reduce the number and strength of policies needed. For example, decades of R&D, coupled with strong policies driving deployment, have helped drive down the costs of zero-carbon electricity generation technologies, including solar photovoltaics and wind turbines. As a result, it is much more cost-effective today to provide zero-carbon electricity than it has ever been in the past. The history of research-based cost declines coupled with well-designed policy shows how R&D fits together with other policy types to drive down costs and accelerate the clean energy transition.

CHAPTER THIRTEEN

Carbon Pricing

Carbon pricing is a critical tool for reducing emissions. Ideally, it should cover all sectors of the economy, although some regions have applied it only to specific sectors. Policymakers can either impose a cap on emissions, using a permit trading scheme to limit emissions and achieve the most cost-effective set of reductions, or impose a tax on emissions. In either case, the price of emitting greenhouse gases will increase, providing an incentive to reduce emissions.

Many economists suggest that carbon pricing, set at the social cost of carbon—the sum of social damages from each ton of carbon emitted—is the only policy needed to efficiently reduce emissions. This assertion is false. However, carbon pricing is an essential part of a policy portfolio for tackling emissions. It influences energy use and investment decisions and, designed well, can encourage a cost-effective set of emission reductions. It can also be used to raise revenue for other policies or technologies.

The impact of carbon pricing depends on its design and on the price. Our modeling of a carbon price set at the social cost of carbon suggests it can deliver at least 26 percent of the emission reductions necessary to meet the two-degree target (Figure 13-1).

Figure 13-1. Potential emission reductions from carbon pricing. (Analysis done using data with permission from the International Institute for Applied Systems Analysis [IIASA]. Data source: Tavoni et al., 2013. Data downloaded from the LIMITS Scenario database hosted at IIASA, https://tntcat.iiasa.ac.at/LIMITSPUBLICDB /dsd?Action=htmlpage&page=about.)

Policy Description and Goal

Most regions of the world do not have a price on carbon emissions. When coal power plants spew CO_2 into the atmosphere or natural gas producers vent methane, they are allowed to do so for free. Yet these emissions come with huge social costs. Carbon pricing creates a cost for producing greenhouse gases, discouraging their emission.

The introduction of a price on carbon emissions reflecting the damage they cause spurs investment in new low- and zero-carbon technologies while encouraging manufacturers, power plants, and others to reduce emissions.

Although it can be a source of cost-effective emission reductions, carbon pricing should not be viewed as a silver bullet. Political barriers tend to result in prices on emissions that are far lower than the true harm they cause to society, resulting in too few emission reductions. Additionally, many important emission reduction opportunities are not price responsive (as discussed in Chapter 2) and therefore will not respond significantly to a price on carbon. An optimal policy package will combine carbon pricing with other policies to capture the price-resistant options.

Carbon pricing policies introduce a cost on carbon emissions either directly or indirectly:

- A **carbon tax** is levied directly through a per-unit charge on emissions of CO_2 (or CO_2e).
- A **carbon cap**, often called a cap-and-trade program or emission trading system, indirectly prices carbon through a requirement that large emitters acquire carbon allowances, each of which entitles the holder to emit a ton of carbon dioxide. The cap limits the number of allowances under the system. A market for allowances enables trading within and between covered sectors.

The core goal of pricing carbon is to reduce emissions, but there are often ancillary goals—equity, reducing conventional pollutants, stimulating economic or technological development, and cutting other taxes, to name a few. The different goals should be made explicit, because they will affect choices about how to design the policy. Of course, any design must be politically palatable, which often requires making tradeoffs.

Pricing carbon to control emissions has several advantages:

- It is technology neutral.
- It is a source of low-cost reductions.
- It creates an incentive that radiates across all sectors of the economy.
- It reduces the information burden on regulators, helping to overcome information asymmetries. Because of the economy-wide nature of carbon emissions and the complexity of modern economies, each firm knows its own operations and market better than regulators ever could. Carbon pricing uses market forces to encourage companies to put that knowledge to work discovering emission reductions.
- The price signal affects investment decisions and behavior for existing energy infrastructure for both consumers and businesses.
- It generates revenue, which can be used to:
  - Reduce inefficient taxes, thus stimulating the economy.
  - Support a faster, more efficient clean energy transition, through deployment incentives (which help bring down future costs and can fund reductions the carbon price itself would not induce) and through research and development (R&D), which helps achieve a broader set of low-cost, low-carbon options.
  - Accomplish equity goals, such as ensuring that disadvantaged communities share in the gains, through emission reduction projects in the most polluted neighborhoods or targeted economic support.

A carbon tax and a carbon cap, in their pure forms, offer a choice between price certainty and quantity certainty. A carbon tax has a fixed price for each unit of emissions, granting price certainty, and the emission outcome that follows is determined by the response of people and businesses to that price. A carbon cap, by contrast, sets a cap on the allowable quantity of emissions, providing quantity certainty, and the price of emissions is revealed through the trade of allowances.

Either a well-designed carbon tax or cap can successfully and cost-effectively reduce carbon emissions. Where there is significant momentum for either a carbon tax or cap, proper design is more important than the type of policy selected; the two approaches are more similar than different. Whether or not they produce a good result depends largely on design and execution.

One of the biggest challenges in designing carbon pricing is finding a way to lower emissions while minimizing economic impact. The optimal policy will achieve steep emission reductions without forcing industry and its associated emissions and jobs to relocate outside the regulated region. Such relocation to other regions is commonly called leakage. The potential for leakage is often overstated, but it can be significant in sectors that rely heavily on global trade. When carbon pricing is applied to these sectors, it can be easy for manufacturers to relocate to other unregulated regions or for foreign manufacturers to outcompete the priced industry. The hybrid approach of blending a carbon tax with a carbon cap, discussed in the next section, helps balance environmental and economic goals.

The guidelines discussed in this chapter should be tailored to each jurisdiction. The following questions can help identify how the guidelines should be modified:

- How developed (and how large) is the economy that will be covered by the carbon price?
- What complementary policies does the jurisdiction have in place, and how will these affect future emissions?
- How strongly have neighboring countries or jurisdictions tackled climate change?
- Do trading partners have a carbon pricing policy or equivalent controls in place or under development?[1]
- Are the policies applied piecemeal, are they customized for different industries, or are they uniform?

When to Apply This Policy

Every climate policy portfolio should include carbon pricing. In addition to being a strong policy for driving down emissions, carbon pricing can overcome some drawbacks of other types of policy while providing new benefits.

For example, it can often be difficult for government regulators to gain access to proprietary information from polluting industries, but carbon pricing helps overcome this information asymmetry. By setting a price and then allowing industries to reduce emissions based on their unique costs, technology options, and emission profile, carbon pricing significantly reduces the amount of information regulators need compared with performance standards or other policy types.

Carbon pricing can also yield public finance benefits. For example, taxing a social harm such as pollution is more economically rational than taxing desirable economic inputs, such as labor and equipment.

Nevertheless, carbon pricing is not a silver bullet: It is a necessary but not sufficient part of an optimal policy portfolio. The optimal policy portfolio instead combines carbon pricing with a broad set of other policy types including performance standards, support for R&D, and supporting policies. Chapters 2 and 3 provide more detailed information on how carbon pricing fits within a broader climate policy portfolio.

Detailed Policy Design Recommendations

In this section, we discuss why creating a hybrid carbon pricing program is important and how to design such a program. The discussion begins by examining emissions and price uncertainty under both a cap and a tax and why this uncertainty implies that a hybrid program is best. After this discussion, we introduce the broader set of policy design principles that apply to carbon pricing.

Why and How to Create a Hybrid Carbon Price Program

Policymakers should consider both the price of carbon (which affects total program costs) and the targeted amount of emission reductions from the program. Because a carbon tax focuses solely on the price of carbon whereas a carbon cap focuses solely on the amount of emission reductions, using a strict version of either of these options can result in prices or emissions that differ from policy goals or expected outcomes of the program.

For example, if a carbon cap is set too low, it may result in higher costs than are expected or tolerable. Similarly, if a carbon tax is set too low, it may not deliver the emission reductions policymakers intend.

Policymakers can address these limitations by combining a carbon tax and a carbon cap into a hybrid carbon pricing program, providing a more balanced approach to managing uncertainty.

Technical Foundation

Here, we introduce the types of uncertainty under either a strict carbon tax or a strict carbon price.[2] The concepts addressed throughout this discussion are illustrated in Figure 13-2.

Figure 13-2 shows an abatement supply curve, which provides information on the cost to reduce the next unit of emissions at a given starting level of emissions. The vertical axis shows the price to reduce a unit of emissions, and the

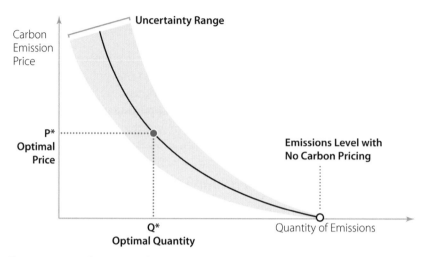

Figure 13-2. A carbon pricing abatement curve.

horizontal axis shows the total amount of emissions. The bolded center curve shows the cost of reducing emissions relative to the total amount of emissions. As total emissions decrease (moving from right to left), eliminating the next set of emissions (moving left again) becomes more expensive. This pattern is also reflected in the marginal abatement and policy cost curves discussed earlier (see Chapter 3).

The abatement supply curve shown in Figure 13-2 can be used to illustrate the expected emission reductions under a carbon tax or the expected price of allowances under a carbon cap.

For example, a carbon tax set at the optimal price (P*) will lower emissions from the level with no pricing to the optimal quantity (Q*). Because reducing another unit of emissions is more expensive than just paying the tax, companies will choose to just pay the tax.

Under a carbon cap, a cap set at the optimal quantity (Q*) will result in the optimal price (P*).

Of course, the price of reducing emissions in the future is somewhat uncertain. For example, technological breakthroughs could significantly reduce the costs of reducing emissions. Similarly, changes in the price of goods or energy might make it more or less expensive to reduce emissions. This uncertainty is captured by the blue area around the black curve.

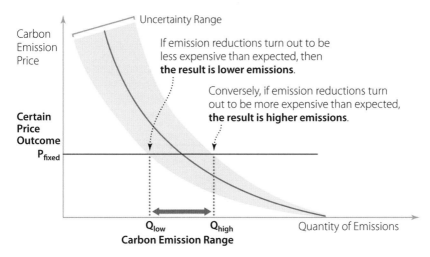

Figure 13-3. At a fixed price, carbon pricing has uncertain emission outcomes.

A carbon tax and a carbon cap handle this uncertainty differently.

Figure 13-3 shows how the total level of emissions could vary under a strict carbon tax.

When the price of carbon is fixed, emissions can be higher or lower than expected, given uncertainty in the cost to reduce emissions. If reducing emissions turns out to be more expensive than expected, emissions might actually decrease only to Q_{high}. Conversely, if emission reductions turn out be cheaper than expected, emissions might fall all the way to Q_{low}. In sum, although a carbon tax provides price certainty, it is accompanied by emission uncertainty.

The inverse is true for a carbon cap, as shown in Figure 13-4.

When the total quantity of allowable emissions is fixed, the cost of reducing emissions to that target can be higher or lower than expected, given uncertainty in the cost to reduce emissions. If it turns out to be more expensive to reduce emissions than expected, the price of allowances under a carbon cap might be as high as P_{high}. On the other hand, if it ends up being cheaper to reduce emissions than originally expected, the price of allowances might fall to P_{low}.

Ultimately, higher-than-expected emissions or higher-than-expected carbon prices might be unacceptable. Fortunately, combining a cap with a tax can balance these risks to reduce uncertainty.

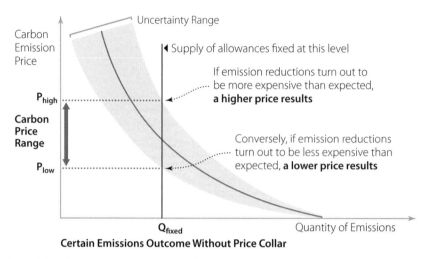

Figure 13-4. At a fixed cap, carbon pricing has uncertain price outcomes.

How to Create a Hybrid Carbon Pricing Program

The ideal carbon pricing program is a hybrid of a cap and a tax. The hybrid approach is built on a carbon cap, with a strong science-based emission target and with a minimum (floor) and maximum (ceiling) permit price. If the price of the carbon cap drops to the floor or hits the ceiling, it becomes fixed at the floor or ceiling price, essentially becoming a tax. This bounding of prices in a carbon cap program is also called a price collar. The use of a price collar helps ensure that a carbon goal will be reached unless carbon prices rise too much. (It's worth noting that carbon reductions the world over have generally been far cheaper than expected.) The hybrid approach turns the weakness of price uncertainty under a carbon cap into an advantage. The price floor creates the potential to drive significant carbon reductions if the price to reduce emissions turns out to be lower than expected, and the price ceiling ensures costs do not rise above acceptable levels. The design of a price collar is discussed in more detail later.

Policymakers create a price ceiling by agreeing to release (usually through an auction) additional allowances if prices reach the price ceiling. When the number of allowances increases, the price of each allowance drops. The option to release additional allowances is sometimes called adding a safety valve: If prices rise too high, policymakers can open this "safety valve" to increase the

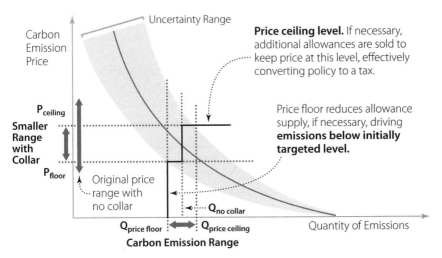

Figure 13-5. A price collar can reduce the price uncertainty of a carbon cap.

number of allowances. This approach effectively caps the price of allowances at the ceiling level, providing the price certainty guaranteed by a carbon tax.

To create a price floor, policymakers establish a minimum allowance price. The price floor allows policymakers to ensure a minimum price on carbon and corresponding revenue even if emission reductions turn out to be cheaper than expected. Over time, the minimum price should rise at a set percentage plus the rate of inflation.

Figure 13-5 shows how using a price collar to create a hybrid carbon pricing program bounds uncertainty.

With the price collar in place, the price is guaranteed to be at least as high as P_{floor} but no higher than $P_{ceiling}$. Compare this range of prices, indicated by the vertical red arrow along the y-axis, with the original range of prices, represented by the blue arrow. The price collar significantly reduces the price uncertainty associated with the carbon cap.

At the same time, the price collar introduces some uncertainty into the quantity of emission reductions, as shown by the horizontal red arrow along the x-axis. The price floor effectively reduces the number of allowances available, ensuring at least a minimum level of emission reductions are achieved. At the other side of the price collar, this program design makes additional allowances available at the price ceiling to ensure prices remain acceptable.

So far, we have looked at how to create a hybrid carbon pricing system starting with a carbon cap. However, policymakers can also create a hybrid program starting with a carbon tax. To create a hybrid carbon tax, the policy must adjust prices based on whether it is achieving the desired emission reductions. If emission targets are being met, the carbon tax does not need to increase. If emission reductions are falling short of desired levels, an increase in carbon tax would go into effect. To date, there are no real-world applications of this approach. However, one such proposal is being discussed in Washington State,[3] and two recent research papers have taken on the topic.[4]

Policymakers should use careful quantitative analysis of policy impacts to identify a desired emission trajectory and price path when determining price floors and ceilings. These analyses should incorporate social costs and benefits, including the value of improved air quality for public health. If policymakers are unable to afford sophisticated modeling, one option would be to borrow approaches of neighboring jurisdictions.

Admittedly, hybrid cap and tax programs are novel, and choices about caps and price floors and ceilings are nuanced. Some impacts will need to be considered on a qualitative basis in parallel with quantitative analysis.

Once the desired price and emission reductions are identified, policymakers must identify the price ceiling, floor, and trajectory.

For a hybrid carbon cap, policymakers should set a price floor below and ceiling above the expected prices from the desired emission pathway. Ideally, the price floor should be set at 50 percent below the expected allowance price in a given year, and the price ceiling should be set at 50 percent above the expected allowance price. The floor and ceiling should rise steadily over time, at a rate of about 5 percent per year, plus the rate of inflation.

For a hybrid carbon tax, policymakers should set a price escalation schedule corresponding to the share of reductions achieved relative to the goal. Emission reduction should be tied to emissions in the start year of the program rather than trajectories of business-as-usual emissions, which are often highly uncertain, if not completely inaccurate. The tax should be raised based on how much emissions fall in response to the carbon price. If emissions do not fall at all, the tax should be increased by 7 percent. If emissions are reduced by less than half of the way to the target, policymakers can try increasing the tax by 5 percent. If emissions are reduced by more than half but still less than the goal, policymakers can try increasing the tax by 1 percent. These price increases should be in addition to the annual inflation rate.

Policy Design Principles
The policy design principles discussed in this section apply to carbon pricing programs.

Create a Long-Term Goal and Provide Business Certainty
Carbon pricing policy should be defined at least a decade into the future. Major investments involve a long planning horizon, and investors need sufficient long-term certainty and commitment in order to integrate a carbon price, or any policy, into the investment calculation.

Build In Continuous Improvement
A carbon price or cap should steadily improve, with prices increasing and emissions falling. In combination with long-term goals, steady improvement builds demand for innovation, helping drive a market for new low-carbon technologies. Allowing the policy to plateau or weaken undermines the ability of companies to confidently invest in R&D.

Capture 100 Percent of the Market and Go Upstream or to a Pinch Point When Possible
A cross-sector policy such as carbon pricing is well suited for broad coverage, which should aim to capture as close to 100 percent of emissions in the market as possible. Broad coverage discourages leakage. For example, if electricity is a covered energy source but natural gas combustion is not, then producers may shift to using natural gas instead of electricity to avoid the cost of a carbon price.

Administrative effort can be reduced by regulating at pinch points—places where the energy flows are concentrated. For example, petroleum refineries, where crude oil is delivered, or the terminal rack, where processed liquid fuels transition from ships, trains, and pipelines to local distribution, are a good choice for capturing oil-related emissions. In the electricity sector, utilities or power generators are better options for the point of regulation than end-use customers.

Carbon pricing programs often target emissions at large industrial sites, which include many types of facilities: petroleum refineries, electric power generators, cement plants, and so on. These facilities vary in size and emissions. Because some of these facilities have very low emissions and the cost of regulating them might outweigh the benefit, policymakers often need to establish a threshold below which emissions are not covered.

Yet the possibility of such a threshold may encourage some emitters to drop below in order to avoid regulation. To avoid industries clustering right below minimum thresholds, regulators can require the reporting of emissions below the threshold level for the policy. For example, policymakers might tax industrial facilities only if they have emissions of 25,000 tons or more, but they can also require reporting of emissions for any facilities with more than 10,000 tons. Setting the lower reporting requirement allows regulators to see whether facilities are clustering right below the coverage threshold.

Prevent Gaming via Simplicity and Avoiding Loopholes
Carbon caps can be subject to gaming because of the nature of trading allowances and the use of offsets, which industries can use to show they are complying with a cap through investments outside the regulated region. Policymakers must establish a central registry to ensure that allowances are not used more than once.

Additionally, if there are only a few companies holding allowances, it creates the opportunity to distort the market and raise the price of allowances. Policymakers should establish limits on how many allowances any single market participant can possess to avoid a concentration of market power that might allow price manipulation. A common foundation for monitoring and enforcement of market power is third-party verification of self-prepared emission reports. Third-party verifiers should be assigned by policymakers rather than left to covered industries.

Additional Design Considerations

Leakage Control
Leakage, discussed earlier, occurs when introduction of a policy encourages the industries being covered to move outside the boundaries of the area covered by the policy. Although many economists believe that leakage due to carbon pricing will be small, it is politically fraught.[5] Much of the concern about leakage has focused on energy-intensive industries highly dependent on global trade, whose costs could increase significantly and make them uncompetitive relative to other unregulated competitors.

Leakage is more of a concern when carbon pricing is implemented in a smaller area, such as a single state or province, because emitters can move a short distance to escape the policy. As more and more regions take meaningful steps to control carbon emissions, the threat of leakage diminishes.

Carbon tax programs typically address leakage by exempting energy-intensive, trade-exposed businesses from having to pay the tax. Carbon cap programs typically address leakage by providing some free allowances to the same set of industries.

Identifying these industries and how many allowances to allocate for free can be challenging. The current state of the art is to use a method called output-based free allocation. "This method offers firms free allowances as a function of their levels of production in the current or in a recent time period."[6]

Where regions are part of a larger electricity system, carbon pricing must cover imported electricity, because it is easy for electricity generation to move outside the covered area.

Auction Most or All of the Carbon Allowances under a Cap

Policymakers have two options for allocating allowances under a carbon cap: giving them out for free or auctioning them off. Auctioning has many advantages over free distribution: It is the simplest approach for government, it avoids the economic distortions that plague all forms of free allocation, it facilitates price discovery, and it is a source of public funds.

Auctions are also the simplest means of introducing a price collar. A price floor is achieved by refusing to accept bids for allowances below a minimum price. A price ceiling is accomplished by making additional allowances available for auction once a price cap is hit. In this way, prices will rise no higher. Auctioning can also avoid turning free allowances into windfall profits for regulated industries.[7]

Free allocation does not suppress permit prices or protect consumers in most cases. Even when allowances are given away for free, there is an opportunity cost to using an allowance, because using it means it cannot be sold. This implicit cost is passed along to customers even under free allocation, unless there is significant competition from firms not covered under the program.

In some cases, a transition period involving some free allocation may be appropriate. Free allocation of some allowances can help protect trade-exposed, energy-intensive industries, which may otherwise be motivated to move to another region with lower energy prices. Over time, however, there should be a transition to 100 percent auction-based distribution, with generic subsidies taking the place of freely allocated permits for trade-exposed industries in need of government support.

Revenue Use

Carbon pricing creates a revenue stream. In turn, the revenue generated through carbon pricing can be used to accelerate the clean energy transition by funding research, development, and demonstration projects or incentives for deployment of low-carbon options that a carbon price alone would not achieve. Using carbon auction or tax proceeds to drive down the cost of carbon abatement through R&D and deployment incentives creates a virtuous cycle, with each increment of carbon abatement helping pave the way for further, cheaper abatement.

Of course, there are many ways in which revenue generated through carbon pricing could be used. Some of the best options include using proceeds for general revenues to reduce distortionary taxes, to pay down the debt, to fund government, or as a "carbon dividend" to be paid directly to the populace; offsetting costs that fall on certain consumers disproportionately, such as low-income households or trade-exposed industries; spurring a virtuous cycle of reducing the cost of carbon abatement by funding R&D and other projects to reduce carbon pollution; subsidizing end-use efficiency, such as building equipment and appliances, a particularly strong candidate because it can at once reduce transition costs for disproportionately affected groups and offer very cheap CO_2 reductions; and promoting equity goals.

Regarding equity, programs should be structured so they do not exacerbate social, economic, or environmental inequities. In fact, intelligent use of the revenue generated by carbon pricing can and should deliver benefits in polluted or disadvantaged communities. Complementing pricing with other more direct pollution control policies is one way to ensure local emission improvements. Another approach is to require some minimum level of investment of revenue in projects with local emission reduction benefits in the most polluted neighborhoods. Both should be core elements of a carbon pricing program.

Allow Banking and Borrowing as Extra Flexibility Options under a Cap

Carbon pricing is inherently flexible, allowing companies and households the choice of purchasing any emission-reducing technology or instead paying to emit. Banking and borrowing offer additional flexibility under a carbon cap.

With banking, covered entities retain unused allowances for future compliance periods. Under borrowing, allowances from future caps can be brought forward and used in present years under some circumstances. Banking and borrowing allow industries to smooth out compliance over time, for example in response to changes in annual availability of hydroelectric power that may follow rainfall fluctuations. Policymakers should limit borrowing to situations in which carbon prices are high and when the reserve to maintain price stability (the safety valve mentioned earlier) has been depleted. This ensures continuous improvement effort as long as prices are manageable.

Linking Carbon Pricing Systems across Different Jurisdictions

Linking carbon pricing systems can deliver greater reductions at lower prices. It makes most sense between programs of roughly equal rigor—allowing "excess reductions" in, say, Quebec to make up for insufficient California reductions. Climate change is a global problem, so more places taking more action is a desired outcome, and linking cap-and-trade programs expands the boundary to find the lowest-cost solutions. Other benefits of linking include creating a "race to the top" by having a state like California set stringent requirements for other jurisdictions' programs as the "price of admission" to link; enabling leaders to coordinate action, which counters notions of unilateral attempts to solve a global problem; and enabling smaller jurisdictions to access a market large and liquid enough to make it worth having a program (i.e., without linking, the jurisdiction would not adopt its own carbon limit).

However, linked carbon cap programs are only as strong as their weakest link, so policymakers should still exercise caution when evaluating whether to link with other regions.[8] Linking will lower the price of permits and will reduce the demand signal that may be needed to successfully bring new, low-carbon technologies to the market. Where there are meaningful differences in environmental stringency, linking does not make sense. Additionally, setting up a program with multiple jurisdictions can increase the challenges of effective governance.

Decision making is also more challenging. The timing and priorities of political decisions in different jurisdictions will always vary, so moving from one target to the next in linked systems is fraught with uncertainty. One way around this is to set an improvement rate (e.g., 4 percent reduction in allowances per year) rather than a specified numerical target some years hence.

Carbon Offsets: If an Offset Program Is Established, Include Strict Protocols and Independent Third-Party Verification

Offsets allow industry to comply with a cap by investing in emission-reducing projects outside the covered region. For example, a cement manufacturer in California might choose to invest in a reforestation project in Colorado as a way to meet its carbon cap requirement, thus offsetting required emission reductions. Offsets can help moderate allowance prices by expanding the reach of carbon pricing policy to projects in sectors that are difficult to directly and completely cover under cap-and-trade, such as agricultural emissions, carbon sequestration, and non-CO_2 gases. Building a serious and effective carbon offset program is complex, and it should be approached incrementally and carefully. To ensure environmental integrity, it is essential to apply strict protocols and independent third-party verification, each reviewed and approved by a public oversight body.

Third-party verifiers should be assigned to projects and paid out of a pool of funds collected from project developers. Verifiers should not be chosen by developers, to avoid creating a dependent relationship, as has been observed in some projects.[9] No offset system has yet been set up with such a strong, independent verification system. A second-best approach is to require periodic rotation of project verifiers and assumption by buyers of the liability, should verification result in offsets being ruled invalid.

An offset program should provide a list of preapproved project types but also allow bottom-up development of new protocols, which can be reviewed, refined, and approved by the supervisory body. Offset options must be periodically reassessed by an expert body to judge whether the project type or technology performance has become common practice and therefore is no longer "additional." A sectoral approach to offsets, where evaluation occurs on a sectoral rather than a project basis, is a promising approach for reducing transaction costs while increasing environmental integrity.

Pros and Cons of Each Approach

Carbon taxes and caps share some strengths and weaknesses: They are equally effective at broad, multisector coverage. Either can provide long-term certainty and opportunities for continuous improvement. The point of regulation for the tax or the allowance can be the same as well. Creating a hybrid program reduces the amount of uncertainty in either prices or emission reductions.

Table 13-1
Pros and Cons of a Carbon Tax versus a Cap-and-Trade Program

Metric	Carbon Tax	Cap-and-Trade Program	Best Option
Environmental effectiveness	May result in higher emissions than intended.	Provides greater emission certainty.	Cap-and-trade
Economic efficiency	Predictable prices support investment and minimize economic disruption.	Less predictable prices are worse for economic performance.	Carbon tax
Fairness	Higher efficiency improves socioeconomic equity.	With more predictable emissions, better for intergenerational equity and environmental justice concerns.	Situationally dependent
Driving technology innovation	Price stability provides market certainty and encourages investment.	Cap can allow higher stringency and resulting carbon price, which could drive more innovation.	Situationally dependent
Linking to other regions	Political economy hurdles limit carbon tax linkage.	Often linked with other regions through cooperative agreements between jurisdictions.	Cap-and-trade
Simplicity	Simple to implement.	Permit system and allocations can be complex.	Carbon tax

Innovative design can further increase similarities. For example, although offsets have traditionally been part of carbon cap programs, some innovative carbon tax programs have allowed the use of offsets.

This section compares carbon taxes to carbon caps without any hybrid adjustments. Table 13-1 offers a high-level summary of the pros and cons of carbon caps compared with carbon taxes.

Environmental Effectiveness

Because they are designed to achieve a specific emission target, carbon caps are generally more effective at reducing emissions to meet a specific target. A carbon cap offers a particular advantage for policymakers looking to develop plans to hit a particular emission reduction target, such as those embedded in many of the international commitments associated with the Paris Agreement on climate change.

Efficiency
Price uncertainty can inhibit both investments in new clean tech ventures and the incumbent fossil fuel combusting industry. Because a carbon price eliminates price uncertainty, it can provide superior economic efficiency.

Fairness
The fairness of impacts, environmental and socioeconomic, will depend largely on design details, with revenue use being a crucial determinant.

Simplicity
In practice, carbon caps are more complicated to implement than carbon taxes. Often, the increased complexity of a carbon cap is caused by the need to determine rules for some free allocation of allowances to energy-intensive, trade-exposed industries. Typically, carbon tax policies have handled the concerns of these industries by allowing them not to pay the tax. Similarly, the inclusion of offsets under a carbon cap can significantly increase its complexity. With full auctioning and no offsets, cap-and-trade is similarly complicated to a carbon tax.

Case Studies

Global Overview
The most obvious pitfall in pricing carbon, worldwide, has been inadequate ambition in setting targets for emission reductions. In the effort to find the balance between environmental stringency and cost containment, policymakers have leaned toward keeping costs low. The social cost of carbon, representing the damage caused by carbon emissions, can be thought of as a reasonable target price. In practice, very few carbon pricing efforts have even approached, much less surpassed, the social cost of carbon. This is not due only to policy challenges, of course: A fortunate side note in carbon pricing programs is that it has been very cheap to hit targets.

For many years, the notion of any carbon pricing at all seemed like a pipe dream. In the 1990s, the elegantly named *Tax Waste, Not Work*[10] and other efforts promoted the concept but with no concrete success. In 2005, the European

Union's Emissions Trading System began operation, and it remains the largest system in the world, covering about 1.8 billion tons of annual carbon emissions. Not long after, in 2009, a group of northeastern U.S. states came together to form the Regional Greenhouse Gas Initiative to cover the region's electric power plants. In 2007, California began planning its carbon cap program, which launched in 2013, and linked with the Canadian province of Quebec in 2014.

A comprehensive survey by the World Bank finds that about 40 nations and two dozen subnational jurisdictions have established a price on carbon.[11] These instruments currently cover about 12 percent of global greenhouse gas emissions. Roughly two-thirds of the coverage, about 8 percent of global emissions, is under a carbon cap, and about 4 percent is subject to a carbon tax. A particularly anticipated development is the expansion of China's pilot projects, which should cover more than 1.2 billion tons of emissions. China's national cap-and-trade program is expected to launch in 2018.

Figure 13-6 aggregates all the world's current carbon prices into a curve, also showing the amount of CO_2 tons covered.

The width of each segment of the stairstep line shows the amount of covered emissions, and its height indicates the price. But for a tiny slice of emissions in Nordic countries, nearly the entirety of the globe's carbon pricing falls well below the U.S. Environmental Protection Agency's social cost of carbon, about $41 per ton.[12] Figure 13-6 also highlights the recommendation of the High-Level Commission on Carbon Prices that countries aim for carbon prices of $40–$80 per ton in 2020 in order to meet the emission reductions agreed to in the Paris Agreement.

Although the European Union's Emissions Trading System has become an accepted part of doing business in Europe, it stands as a cautionary tale about the potential for a very large bank of allowances to accumulate and cause persistently low allowance prices (the program has no auction price floor). The price has varied from €3 to €10 since 2011, standing at €7 per European Union Allowance as of October 2017. The onset of the financial crisis in late 2008 caused a fall in emissions due to reduced economic activity. Other renewable energy and energy efficiency policies also drove emissions down. The result was an oversupply of allowances that reached more than 2 billion tons in 2013, at the start of the program's third compliance period. Policymakers are tackling the problem by taking some allowances out of future caps and delaying some auctions. Meanwhile, political barriers have impeded adoption of a price floor.

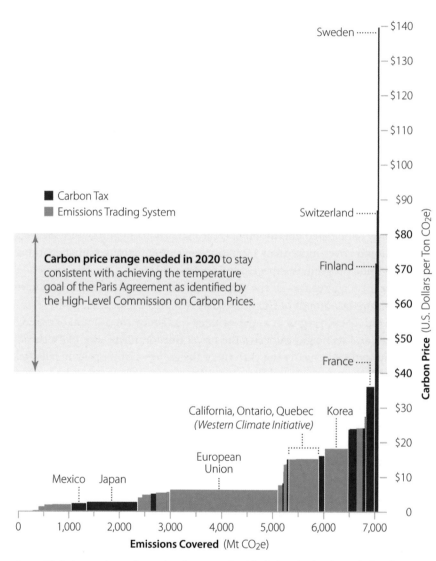

Figure 13-6. In practice, carbon prices have remained far below the levels needed to drive enough emission reductions. (World Bank; Ecofys; Vivid Economics, 2017. *State and Trends of Carbon Pricing 2017*. Washington, DC: World Bank. © World Bank. https://openknowledge. worldbank.org/handle/10986/28510 under Creative Commons Attribution 3.0 IGO license.)

Carbon taxes have been most robustly used in Nordic countries, where prices range from about $25 per ton in Denmark to roughly $50 per ton in Norway and Finland and $130 per ton in Sweden. These Nordic countries have mostly used new government revenue to lower taxes on labor. Japan greatly expanded the emissions covered by a carbon tax when it introduced one in 2012, but at less than $2 per ton it provides a weak incentive. Canada's new commitments around carbon pricing should be a game changer in stringency for taxes, with a carbon price of $10 in 2018 and increasing to $50 in 2022.

Regional Greenhouse Gas Initiative Linked Carbon Cap Program

The Regional Greenhouse Gas Initiative covers the electricity sector CO_2 emissions in nine eastern American states. A key highlight is the program's early embrace of auctioning as the main method of distributing allowances. The Regional Greenhouse Gas Initiative was the first program to fully auction allowances, illustrating the economic benefits that can be created by smart investment of auction revenue. Revenue generated by the Regional Greenhouse Gas Initiative has funded energy efficiency improvements, which have created an array of economic benefits, starting with consumer savings of more than $618 million,[13] and spending of extra disposable income from energy efficiency and local clean energy investments has generated more than $2.9 billion in additional economic growth.[14] Public health benefits worth $5.7 billion are estimated to have come from reductions in fine particles and smog-causing emission, which are co-benefits of lowered carbon emissions.[15]

Time and again, modeling in advance of cap setting has resulted in business-as-usual emissions that are higher than the real-world result, and the Regional Greenhouse Gas Initiative is an example of the problems with using this approach.[16] As a result of basing allowances on a forecasted emission scenario, the Regional Greenhouse Gas Initiative program has wrestled with overallocation. Despite the governance challenges in a linked system, the Regional Greenhouse Gas Initiative system has been regularly tightened to deal with this oversupply. In addition to the cap adjustments discussed later, the Regional Greenhouse Gas Initiative has established a regular 4-year program review and recalibration process.

In 2005, when the cap was set, natural gas prices were high and rising, as were emissions. The intent of the states was to set the cap at the expected levels in 2009, keep it flat for 5 years (when emissions were otherwise expected to continue to grow), and then decrease it by 2 percent per year through 2019.

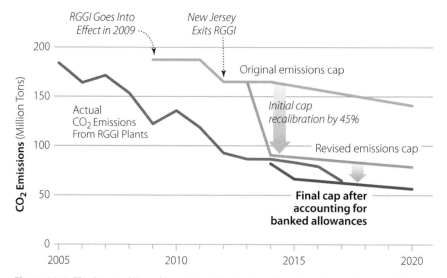

Figure 13-7. The Regional Greenhouse Gas Initiative has adjusted cap levels downward in response to new information about program performance. ("Regional Greenhouse Gas Initiative Auction Prices Are the Lowest since 2014," U.S. Energy Information Administration, 2017, https://www.eia.gov/todayinenergy/detail.php?id=31432.)

It was a surprise when gas prices fell precipitously in the intervening years, displacing a great deal of coal and causing emissions to fall far below the cap before the program had even launched, as shown in Figure 13-7.

Figure 13-7 shows that the initiative was more than 50 million tons oversupplied from the outset. By 2012, emissions had fallen to about 90 million tons, about half the level of the 180-million-ton cap. Participating states made two adjustments to the cap in response to the oversupply of permits. First, they lowered the cap to account for the misjudgment of underlying technological trends; modelers had not anticipated the decline of coal with the emergence of natural gas as a cheaper alternative. In a second adjustment, the states lowered caps by 140 million tons to account for excess allowances sold and banked from 2009 to 2013.

In 2017, the program completed another review, tightening the post-2020 cap, which had previously been set to flatten. Under new commitments, the cap will provide an additional 30 percent reduction in emissions by the year 2030 relative to 2020 levels.[17] As part of this adjustment, the Regional Greenhouse Gas Initiative will also undertake another adjustment for banking.[18]

The price floor included in the initiative has saved the program's allowance price from completely collapsing. By 2017, the price floor had reached $2.15 per ton,[19] with an escalation rate of 2.5 percent per year. Thus, the price floor ensured the program provided some revenue, although it did not truly solve the problem of overallocation. At such low prices, it costs little to purchase allowances as a hedge against future higher prices.

The Regional Greenhouse Gas Initiative includes a soft price ceiling by allowing a set of reserved allowances to be released if the price hits certain levels: $4 in 2014, $6 in 2015, $8 in 2016, and $10 in 2017, rising by 2.5 percent each year thereafter. Prices have remained low, however, peaking at $7.50 per ton at the December 2015 auction. However, in light of the low soft price ceiling, some release of reserve allowances has happened.

In sum, the Regional Greenhouse Gas Initiative demonstrates how policymakers can adapt to oversupply, why auctioning of allowances is important, and the value of having a price floor at auction. However, the program suffers from limited coverage (it covers only the power sector), failure to address leakage (it does not cover imported electricity), and an overly weak price collar.

California–Quebec–Ontario Linked Carbon Cap Program

The California–Quebec–Ontario linked cap-and-trade program is the best example of cap-and-trade design. The California and Quebec programs launched separately in 2013 and joined together in 2014. Linkage with Ontario was agreed to in 2017 and took place in 2018. California is the largest emitter among the three, with 62 percent of emissions, compared with 26 percent for Ontario and 12 percent for Quebec.

The program has the widest coverage of any large carbon cap, covering about 80 percent of emissions across the entire economy and almost all fossil fuel combustion. The program accounts for imported electricity, reducing the problem of electricity generation shifting outside the program's borders to avoid the carbon price. Program design is mostly aligned among the jurisdictions, although some differences exist.

The program's most outstanding feature is its price collar. The California price collar started at $10–$40 per ton, and it increases annually at a rate of 5 percent plus the rate of inflation. In 2017, the price collar range was $13.57–$50.70 per ton. This price collar is the highest of any carbon cap.

California has mostly auctioned allowances rather than giving them away in order to prevent windfall profits. In an interesting hybrid allocation approach, state electric utilities receive free allowances but are required to sell them at state auctions in what is called a "consignment auction" approach. The revenue is sent back to privately owned utilities with stipulations. Funds are returned as a lump sum payment per customer, which has the effect of counteracting price increases while also retaining a carbon price signal.

California's design is not without potential for improvement. The California case offers another example of how difficult it has been to achieve sufficient stringency. The program's emissions have fallen below cap levels because of a combination of the success of other policies in driving down emissions and the recession of 2009–2010, leaving emissions lower than originally expected. As a result, the state's 2020 target is turning out to be easier and cheaper to meet than expected.

The carbon cap is successfully playing the role it was given in the context of the package of policies put in place to achieve the 2020 target. That package of policies relied primarily on performance standards and other sector policies. Before implementation, cap-and-trade was expected to deliver only about 20 percent of reductions. In reality, cap-and-trade has driven less than that, but this is not a problem because it was always intended as a backstop.

However, oversupply in the system, if left unaddressed, threatens to undermine the effectiveness of the carbon cap, a significant problem because California is increasingly relying on it as the linchpin of its climate policy efforts.[20]

Oversupply and banking at this level could significantly reduce the future effectiveness of the carbon cap. California's 2030 strategy envisions the carbon cap delivering 40–50 percent of emission reductions from 2021 to 2030, in the range of 240–300 million metric tons. Eventually, the program will need to confront oversupply, probably as the Regional Greenhouse Gas Initiative did, with future caps adjusted downward to account for banked allowances from oversupply in its first years.

In sum, the California–Quebec–Ontario cap-and-trade system is exceptional for its broad coverage, high price floor, and consignment auction innovation. However, as with other carbon caps, oversupply is a serious concern.

Conclusion

Carbon pricing is not a silver bullet to achieve the deep emission reductions needed to meet the two-degree target. Rather, it is one important part of a package of policies such as those outlined in the other chapters of this book.

The steadily increasing adoption of carbon pricing attests to the positive real-world experience so far. There have been no major breakdowns or market disruptions that might have led to a loss of confidence in the approach. The policy has proven to be an attractive source of revenue. Economists and public finance experts universally agree there are efficiencies to raising funds through charges on pollution or other socially harmful activities. For regulators, carbon pricing has somewhat lower informational demands, providing a cross-sector tool to achieve cost-effective reductions above and beyond other policies.

The major limitation so far is that policymakers have been overly cautious. Taxes have been too modest and caps too generous, evidence of the substantial political hurdles and the limits of state-of-the-art economic and technology forecasting. Our suggestion to focus on proper design instead of whether to use a tax or a cap aims to move the dialogue past clashes of worldviews to practical design considerations. Design of either can be simple or complex, and they have similar enforcement requirements. Each can be structured to counter its weakness through hybridization. Carbon pricing policy should embody scientifically grounded emission targets while using proven mechanisms to keep prices within reasonable bounds. If chosen, carbon taxes should be quantity-adjusted, with prices ratcheting up if emission impacts fail to materialize as expected.

CHAPTER FOURTEEN

Research and Development Policies

Countries cannot be safe, prosperous, and healthy unless they have a broad range of energy technology options. Energy technology can help meet five goals:

- Energy supplies should be affordable.
- Energy delivery should be reliable.
- Energy systems should not unduly harm the environment.
- Energy companies should be competitive and should create good jobs.
- Energy choices should not jeopardize national security.

All these goals are easier to achieve with a steady, strong offering of new technologies and improvements to existing technologies. Advances in the last two decades have opened up vast new reserves of natural gas, have made thermal power plants increasingly efficient and clean, have driven down the cost of solar and wind, and have made it possible to reduce energy consumption in appliances and buildings by 50 to 90 percent.

That's good news. The bad news is that many countries are starving future generations of the next set of options. For example, in the United States energy companies spend less than 0.5 percent of their sales on new technology research and development (R&D).[1] This contrasts with information technology (R&D is 20 times higher as a fraction of sales) and pharmaceuticals (almost 40 times higher).[2] A handful of nations stand out from the rest of the world as they build strong positions on energy R&D: South Korea, Israel, Japan, and three Scandinavian countries make annual R&D investments in excess of 3 percent of GDP.[3] Much of the world lags. If we don't get serious about inventing future energy technologies, energy will become a burden on economic productivity, particulate pollution will continue to cut short millions of human lives, and global warming will worsen.

That said, accelerating technology development without wasting money can be a challenge. Fortunately, there are proven methods that can dramatically increase the rate of success. This chapter describes a handful of best practices that can help energy technologies advance all the way from the laboratory to the marketplace. This work is built on experience in the field, collaboration with government, reviews of a dozen studies, and many interviews with experts from the private sector, academia, and national labs.

Policy Description and Goal

R&D policies aim to improve the effectiveness of government R&D and support R&D investment from the private sector. Countries with strong R&D also stand to benefit from the ability to sell new technologies abroad.

The relationship between investment in R&D and emission reductions is highly uncertain. We therefore do not estimate the potential impacts of R&D on meeting long-term emission targets. Nevertheless, R&D policies can reduce the costs of emission reductions, improve the performance of existing technologies, and unlock new technologies that can ultimately make carbon reduction easier and more cost-effective.

When to Apply This Policy

The starting point in thinking through technology development is to understand that different strategies are needed for different stages of a technology's lifecycle.

R&D policies are most effective for technologies early in their lifecycles. Any country that has either public or private R&D efforts aimed at bringing new technologies to market can benefit from R&D support policies.

A country that does not have substantial resources to invest in public R&D efforts (such as by establishing or funding national labs) may be able to entice multinational companies to conduct R&D locally, if the country can provide favorable tax incentives and a sufficiently strong educational system to provide the highly skilled scientists and engineers necessary for R&D operations.

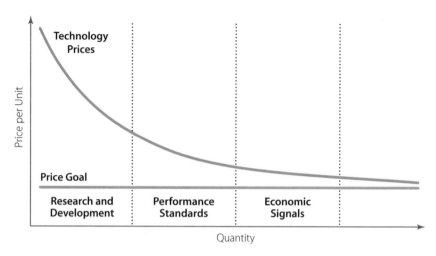

Figure 14-1. The policy–technology learning curve (illustrative).

For example, Ireland is known for proactively courting innovation-focused multinational companies by using these mechanisms.[4] Developing countries without the necessary conditions for either public or private R&D efforts are technology importers, and these countries may be better served by focusing on other types of policy, such as performance standards for vehicles and appliances, policies to promote renewable energy development, and financial assistance for building and industry energy efficiency upgrades.

Even within the R&D span of time in Figure 14-1, conditions are not uniform. Government financial support (whether through grants, loan guarantees, or other mechanisms) is most crucial where gaps exist in other funding opportunities. This gap most commonly happens between basic research (which is often tackled through government labs and academic institutions) and the scaling up of production for commercial use (which is funded by the private sector), as shown in Figure 14-2. This gap is commonly referred to as the "valley of death," because promising technologies often fail to cross the gap and move from the laboratory to full-scale demonstration projects or production.

Some R&D policies help at all stages of technological development, such as ensuring companies have access to high-level science, technology, engineering, and mathematics talent. Others, such as grants or loan guarantees, may be most effective when designed to help technologies overcome funding gaps.

Funding/Investment

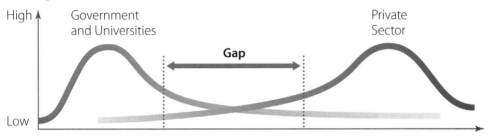

Manufacturing-Innovation Process

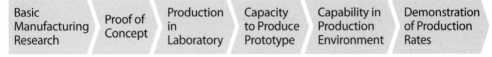

Figure 14-2. The gap in funding availability between basic research and commercialization is typically referred to as the "valley of death." (From "Nanomanufacturing: Emergence and Implications for U.S. Competitiveness, the Environment, and Human Health," U.S. Government Accountability Office, 2014, http://www.gao.gov/assets/670/660592.pdf.)

Detailed Design Recommendations

Research fundamentally involves speculation and experimentation. Even with perfect policy incentives, many research projects will never become commercial products. Sometimes technical or scientific issues interfere, sometimes the marketplace changes, and sometimes a more innovative approach makes a project obsolete before it can be commercialized. This reality demands that policymakers and investors tolerate risk and research failures. If you don't have any failures, then you will not find true successes.

Sometimes government may be rewarded for taking on this risk. For example, the U.S. Department of Energy's loan guarantee program for clean energy achieved a $30 million profit from 2010 to 2013, despite backing several unsuccessful companies.[5]

Create Long-Term Commitments for Research Success

One of the most challenging issues at the interface of legislation and technology development is the need for a long-term outlook for technology policy. Private sector companies need consistency and reliability before they make big bets. Both government and private sector labs must buy equipment, recruit experts, and build and run careful experiments. Policies that promote R&D therefore must match the long time horizons of technology development, or researchers and government will squander opportunities and waste money.

Confronted with political and budgetary challenges, policymakers tend to fund things a year at a time. But it cannot be overemphasized just how deleterious stop-and-start policy is to serious energy innovation. For example, since its introduction in 1981, the U.S. R&D tax credit was repeatedly extended for short periods of time and allowed to expire. One CEO of an especially research-driven energy technology company told us that, as a result, it "considers the R&D tax credits just to be a windfall, with no impact on the company's R&D choices." In 2015, after 15 piecemeal extensions,[6] the R&D tax credit was finally made permanent.[7]

By ensuring that commitments to R&D last for long periods, government, academia, and private companies will have the confidence to rely on that support when making R&D investments.

Use Peer Review to Help Set Research Priorities

Selecting research projects among many competitors is difficult and complex. When prioritizing different research projects to receive government funding, the government should conduct peer review of the options, involving experts from both within the government and within industries that might benefit from technological progress in the relevant field. Consulting with experts inside and outside the government can help ensure projects are technically feasible, would be useful to society if accomplished, and have an acceptable risk–reward profile.

For example, the U.S. Department of Energy developed a "Quadrennial Technology Review"[8] that considered the potential for breakthroughs in many areas and overlaid them with national priorities, such as reducing dependence on imported oil. The work engaged 600 experts from the private sector, national labs, and academia. The experts were asked to consider the technologies' leverage against a list of national policy goals and against three explicit measures of potential:

- **Maturity:** Technologies that have significant technical headroom yet could be demonstrated at commercial scale within a decade.
- **Materiality:** Technologies that could have a consequential impact on

meeting national energy goals in two decades. "Consequential" is defined as roughly 1 percent of primary energy.

- **Market potential:** Technologies that could be expected to be adopted by the relevant markets, understanding that these markets are driven by economics but shaped by public policy.

This process helped the Department of Energy identify issues with its past funding methods (such as the need to achieve a better balance between projects with near-, medium-, and long-term impacts) and helped identify where to focus efforts to better achieve national priorities.

Use Stage-Gating to Shut Down Underperforming Projects

Research is an inherently risky endeavor, and there is always a possibility that a line of inquiry will produce no results or will produce too few results to justify the necessary investment. To ensure large amounts of money and staff time are not wasted, it is important to establish stage-gates or milestones that a research project must pass in order to continue to receive funding. A project should be shut down if it fails to achieve these milestones, so the staff and funding allocated to that project can be reallocated to more fruitful endeavors. Although some research failures are inevitable, a strong gating procedure helps ensure that when you fail, you fail early and fail fast, before vast quantities of money have been expended.

Funded projects can generate entrenched interests, making it more challenging to remove funding from an existing project than to fund a new project. Therefore, it is critical that gating include independent experts with a combination of scientific and industrial expertise in the relevant field. With the inclusion of an industry perspective, project funding decisions can be made based on a project's scientific merits and ultimate commercialization potential, not on political considerations.

For example, the U.S. Department of Energy's Industrial Technologies Program uses stage-gates to manage its R&D allocations.[9] An example review team might be composed of:

- A representative of the funding department
- Outside (i.e., industry) technical expertise
- Inside (i.e., government) technical expertise
- Representatives of proposed end users
- Members of the R&D team

Concentrate R&D by Type or Subject to Build Critical Mass

Providing a small quantity of R&D funding to each of many different institutions is inefficient, because coordination between these institutions and duplication of work will consume an inordinate share of the R&D investment. It is better to concentrate R&D funding on a specific topic into a smaller number of institutions—potentially co-located with each other or with relevant industry players—to reduce coordination challenges, facilitate knowledge sharing, and avoid duplication of work.

One way to accomplish this is to create "innovation hubs" or "centers of excellence," each of which is composed of academic, private sector, and government researchers, ideally in the same metro area. In addition to avoiding duplication of work, bringing together researchers with different backgrounds unlocks further synergies. Researchers feed off each other's ideas, students gain technical skills through internships and university–industry partnerships, and businesses have access to talent. Business interests working side by side with academia make technologies' transitions from lab to market faster and more reliable. Early-stage investors, such as venture capital, move in and work as a further accelerant.

Make High-Quality Public Sector Facilities and Expertise Available to Private Firms

In some countries, such as the United States, the government has invested in the development of extremely expensive, high-quality scientific and engineering research facilities, such as the Department of Energy's National Laboratories. These facilities are staffed with skilled experts. A private company that wants to conduct R&D to improve the performance of its products might be unable to afford to build its own cutting-edge laboratories and staff them with experienced scientists only to make small improvements to its product line. A research partnership with a national lab allows the company to benefit from high-quality facilities and expertise for a comparatively small payment, enabling it to gain the benefits of research without replicating R&D capabilities. A national lab can partner with many different companies, improving technical prowess across the economy (as long as careful attention is paid to protecting intellectual property). These partnerships can also provide a source of revenue for the national lab, making it less dependent on taxpayer funding.

For example, India's Central Power Research Institute has housed R&D facilities for use by government, industry, and utilities alike over the past 50 years. The public–private research facilities have helped India make progress on high-voltage transmission, power system resilience, and other electricity distribution components.[10]

Meanwhile, the U.S. National Renewable Energy Laboratory has built more than a dozen centralized testing facilities—such as the Energy Systems Integration Laboratory, which studies grid modernization. Similarly, Sandia National Laboratories offer 18 test facilities (such as the Combustion Research Facility and the Mechanical Test and Evaluation Facility) where researchers from private companies, academia, other laboratories, and state and local governments may visit and perform research,[11] or they may contract directly with Sandia to perform their testing.

Protect Intellectual Property without Stifling Innovation

Intellectual property (IP) protections are necessary to protect private firms' investments in R&D. If patents are not protected, then any firm can use research results in its own products, reducing or eliminating the incentive for firms to engage in R&D in the first place.

However, it is also important to avoid allowing patent and IP protections to stifle innovation. The difficulty inherent in ensuring the novelty and uniqueness of every patent submission has led to two problems.

First is the rise of patent assertion entities (also known as patent trolls), companies that acquire a key patent (often a vague or overly broad patent) without any intent to use the patent in a product or service. Rather, they sue carefully selected companies that could be construed as violating the patent, in hopes of extracting a legal settlement.[12] This requires small companies to agree to unreasonable settlement demands or to spend a larger amount of money defending themselves against a frivolous lawsuit, providing a disincentive to innovate.[13]

Second, in some fields, such as information technology, existing patents are numerous, and the need to use underlying technologies is so universal (e.g., to achieve interoperability with other systems or hardware components) that avoiding infringement while achieving innovation is impossible. Large technology companies have learned to defend themselves by acquiring thousands of patents and threatening to countersue any company that sues them for patent infringement. Rivals recognize that suing a deep-pocketed company with many patents will probably result in high legal fees and substantial risk to their own product lines and so are discouraged from suing. However, small companies that do not own thousands of patents don't have this deterrent capability, so they are vulnerable to lawsuits that may force them to go out of business or to sell their company to one of the technology giants.[14] This is far removed from the original intent and purpose of a patent system.

Designing appropriate IP protections is complex. Some advocacy groups have thought carefully about patents and devised principles that can be used to develop suitable patent protections.[15]

Ensure Companies Have Access to High-Level STEM Talent

In order for private companies to conduct R&D successfully, they need a ready supply of talented people with skills in science, technology, engineering, and mathematics (STEM). From a public policy perspective, there are two ways that governments can assist.

The first is to establish top-quality education programs focusing on these areas, helping students acquire science and math skills early and providing a route to further develop these skills at the university and graduate school levels. In the United States, primary and secondary schools are funded primarily by state and local governments, so schools in less wealthy communities receive less money and produce students with poorer scores in science and math (as well as other subjects). Accordingly, in addition to directing sufficient resources to STEM, policies to tackle income inequality and poverty can contribute to improving access to technical education.[16]

Government-sponsored research internships at labs or private firms can help students further develop technical skills. For example, the government of Ireland funds university students and postdocs while they work at internships in the R&D divisions of innovative companies such as IBM.[17]

The second policy mechanism is to ensure that immigration laws enable companies to hire skilled technical talent from other countries. Researchers are highly skilled people who contribute to a country's economy. In the United States, groups from across the political spectrum have advocated for streamlining visa and permanent residency procedures, including a proposal to offer automatic residency to graduates of U.S. universities with advanced degrees in the STEM fields.[18] Australia, Canada, and the United Kingdom use point-based or merit-based immigration systems,[19] which give priority to people who possess degrees and work experience in areas of need, which typically include these fields.[20]

Case Studies

The U.S. Advanced Research Projects Agency–Energy

Seeking a highly effective way to fund research into energy technology and bring new technologies from the laboratory to the market, in 2007 the United States established a new R&D funding agency: the Advanced Research Projects Agency–Energy. The agency's approach was modeled after that of the Defense Advanced Research Projects Agency, a hugely successful government R&D operation that played a key role in the development of technologies we use every day, including GPS satellites, packet-switched computer networks, and the internet.[21]

The Advanced Research Projects Agency–Energy focuses on funding research projects that are too early to attract private sector funding (such as venture capital) but that have the potential to rapidly advance and achieve commercialization. Thus, they span the gap between basic research and product development. Funded projects must be transformational: They must have the potential to "radically improve U.S. economic prosperity, national security, and environmental well-being."[22] Projects must have specific, proposed applications toward products or processes that could be commercialized, and the agency provides resources for research teams on how to seek commercial funding and proceed down the path of commercialization to follow up their grants. The Advanced Research Projects Agency–Energy also adopts a nimble funding structure, making funding decisions quickly and relying on program directors who are experts, often from industry, and who "serve for limited terms to ensure a constant infusion of fresh thinking and new perspectives."[23]

The Advanced Research Projects Agency–Energy has distributed $1.5 billion in R&D funding to more than 580 projects since funds were first disbursed in 2009. Many recipients have gone on to form new companies and partnerships with other funding entities.[24] Advances have been achieved in grid-scale and flow batteries, electric vehicle systems, power flow and grid operations, power electronics, advanced materials, and more.[25]

Innovation Network Corporation of Japan

In 2009, Japan launched the Innovation Network Corporation of Japan, a $1.9 billion collaboration between the public and private sectors to achieve advances in energy, infrastructure, and other high-technology sectors. The

Japanese government invested 95 percent of the upfront capital to create the corporation, and 26 private companies made up the final 5 percent investment.[26] The Innovation Network Corporation is an investment company. It directs its investments in order to nurture the development of next-generation industry through applied technology, focusing on innovations that will have "social significance."[27] The company aims to achieve positive economic returns from its investments[28] so it will not need ongoing government support. Example investments in the energy space include small wind power, laminated lithium ion batteries, smart meters, and semitransparent solar cells.[29]

The Japanese government has offered $8.5 billion in loan guarantees for the corporation's investments, mitigating the risk in the event that some investments perform poorly.[30] The Innovation Network Corporation has formed partnerships with 10 external organizations, including Japanese universities, government agencies, venture capital, and several research institutes.[31] This allows it to benefit from the knowledge and talent of other organizations when making investment decisions.

Germany's Fraunhofer-Gesellschaft

The Fraunhofer-Gesellschaft is a network of 69 research institutes throughout Germany. Fraunhofer emphasizes applied research: Most projects last no longer than 2 years "and focus on immediate, applicable results."[32] This helps to fill the budget gap between basic research and commercialization (the "valley of death"). Fraunhofer is the largest research organization in Europe, with a staff of 24,500 and an annual budget of €2.1 billion.[33] Thirty percent of the organizations budget comes from the public sector,[34] and 70 percent is derived from contract research done for public or private entities.[35]

Fraunhofer's institutes are grouped into eight alliances covering specific research areas, such as "materials and components" or "microelectronics." These groups coordinate research, pool resources, and avoid duplication,[36] serving as an example of an "innovation hub" model (as discussed earlier).

Fraunhofer also plays a role in cultivating technical talent, necessary to ensure that German companies and Fraunhofer itself have access to the scientists and engineers they need for R&D success. Each Fraunhofer institute is partnered with a university, and Fraunhofer employs graduate students and postdocs part-time, helping them acquire industry experience alongside their academic studies. "Graduates typically spend from three to six years at Fraunhofer before moving on to positions in industry or academia."[37]

Fraunhofer has played a role in ensuring that German manufacturing businesses remain globally competitive, even in the face of low-cost products from Asia. Many German small and medium enterprises are market leaders for their products, which offer higher quality and performance than inexpensive alternatives. As a result, manufacturing accounts for 21 percent of the German economy, a much larger share than in similarly developed, high-wage countries such as the United States (13 percent) and the United Kingdom (12 percent).[38] Fraunhofer is an example of how policies to strengthen R&D, designed well and applied consistently, can offer outsized returns to a national economy.

Conclusion

A robust pipeline of new energy technologies is crucial to enable a country to continue to meet energy demand and expand the economy while transitioning to clean energy. Strong R&D support policies are crucial early in the lifecycle of a technology, before it reaches market and other policies (such as performance standards and financial incentives) can take over. In particular, many research projects find it hard to acquire funding to overcome the "valley of death," the gap between basic research and early commercialization. This presents an opportunity for particularly high-leverage R&D policy support.

Government R&D financial commitments, programs, and incentives should be guaranteed for the long term, to match the timeframe needed to scope a research project, hire staff, expand or retool laboratories, and turn early research into a marketable product. Peer review, including experts from government and industry, can help identify priorities for research dollars that will achieve practical benefits for the private sector, the environment, and public health. Stage-gating can be used to ensure research dollars and staff time are not wasted, and concentrating research by subject into "innovation hubs" can improve coordination, reduce administrative burden, and accelerate progress. Carefully designed IP protections, as well as education and immigration policies that provide an abundance of top technical talent, lay important groundwork for R&D success in both the public and private sectors.

With smart policies to promote R&D, national governments can strengthen their technological prowess and economic position, attract R&D investment, improve energy security, and reduce pollution.

CHAPTER FIFTEEN

Policies for a Post-2050 World

This book is concerned largely with the design of policies that will be most effective at achieving large emissions cuts before 2050. However, many of the policies discussed in this book (e.g., energy efficiency standards for buildings and vehicles, incentives or requirements for reduced industrial methane emissions, and even carbon pricing) cannot, by themselves, reduce emissions to zero or below.

In the scenario used throughout this book, which gives the world a 50 percent chance of staying below 2°C of warming, emissions become negative after about 2070. This result is common across climate models: To have at least a 50 percent chance of keeping the world below two degrees of warming, global emissions must steadily decline to zero and ultimately become negative (i.e., removing more CO_2 from the air than is added) in the second half of the century.[1] Even if the two-degree target is not achieved, to stabilize the climate at any temperature whatsoever (e.g., 3°, 4°) will probably require similarly dramatic emission cuts, albeit later in time. Therefore, there is no way around the need to reduce emissions to near-zero or below.

This chapter considers technologies that may be necessary in the long term (after 2050) to achieve the emission reductions required by a future with less than two degrees of warming and policies that may be necessary to adapt to climate change. The technologies discussed in this chapter may not be ready for widespread deployment today, but policies to accelerate their progress must begin now so that they will be sufficiently mature by the time they are needed.

One important note: The measures here must not detract from efforts to rapidly cut emissions using the more traditional tools discussed in the rest of this book. Progress on the techniques needed for post-2050 complete decarbonization should be completely additional to and simultaneous with efforts to achieve near-term impact. There is no mopping up the last 10 percent of carbon emissions if we don't eliminate the first 90 percent! When allocating limited resources, it is important to remember that many post-2050 solutions are still in research and development (R&D) stages and therefore require less money to make satisfactory progress in the near term than solutions that are ready to deploy at global scale (e.g., wind, solar, efficient industrial equipment and building components).

Policies to Support Post-2050 Technologies

This section discusses policies to help emerging technologies for achieving zero or negative emissions. Although support for further research in these technologies is broadly needed, three policies in particular can help accelerate their development.

Government support for R&D, discussed in Chapter 14, is the core policy for ensuring that carbon-reducing technologies reach maturity. These technologies have large, positive social externalities (benefits whose economic value cannot be captured by the company deploying the technology), so without government support, companies may choose to direct their R&D efforts elsewhere.

Strong carbon pricing, discussed in Chapter 13, is critical to accelerating these technologies. By putting a price on carbon, governments can help create additional economic value for carbon-reducing technologies and encourage private sector investment.

Finally, some of these technologies will require large-scale demonstration plants or projects to achieve cost reductions through learning by doing. Therefore, governments will probably need to subsidize the construction and operation of a number of demonstration plants or large projects until the technologies discussed in this chapter are better understood.

Technologies for Further Reducing Emissions

Carbon Capture and Sequestration

It may be possible to fully eliminate carbon emissions from the electricity system and to electrify many end uses. For example, renewables and nuclear may be able to supply all electricity needs when combined with flexible demand, large balancing areas, energy storage, and overbuilding wind and solar while putting excess electricity to a useful, non–time-sensitive purpose such as the creation of hydrogen. However, there are some sources of CO_2 emissions that may be difficult to eliminate. For example, manufacturing the clinker in cement releases CO_2 emissions (as described in Chapter 12), and the share of clinker in cement likely cannot be reduced below a certain percentage without affecting the material's structural properties. Another example is the manufacture of new iron and steel (rather than reforging scrap iron and steel in an electric arc furnace), which uses carbon not just as a source of energy but also as a chemical-reducing agent.

Innovations in material science may one day allow the replacement of cement or steel with novel materials with similar structural properties.[2] However, it may not be possible to eliminate all industrial emissions, particularly if efforts to develop and commercialize novel materials encounter problems or cannot be scaled cost-effectively to satisfy the global demand for these materials.

Carbon capture and sequestration (CCS) provides a means whereby humans may continue to manufacture traditional materials without adding CO_2 to the atmosphere. A CCS system extracts CO_2 from a stream of waste gases, uses pressure to liquefy the CO_2, transports it to a geologically suitable region, and pumps it underground for indefinite storage. CCS is already used successfully in the oil and gas industry for enhanced oil recovery, and demonstration facilities using CCS for industrial process and power generation exist around the world.[3] Some CCS power plants might use the Allam cycle, a combustion process that uses CO_2 as the working fluid and produces a very pure stream of CO_2 exhaust, which is easier to capture than CO_2 diluted in air.[4]

CCS may be used by power plants burning biomass (such as wood) rather than coal or natural gas. This is called bioenergy with CCS. Because the carbon in biomass was recently removed from the atmosphere by plants, storing it underground reduces atmospheric CO_2 concentrations.

In addition to the challenges related to the CCS technology itself, bioenergy CCS faces additional hurdles. One issue is the amount of land needed to grow bioenergy crops, which may be very large.[5] Care must be taken to ensure bioenergy CCS does not result in food insecurity or in deforestation to obtain additional cropland. There exist promising research directions that aim to address these challenges. For example, more R&D is needed to develop multifunctional land uses (e.g., to allow the same land to produce food and bioenergy crops). Another route is to derive high-value alternative fuels from bioenergy crops (e.g., liquid transportation fuels) before the residue is burned for bioenergy CCS, thereby improving the economics of devoting land to bioenergy crops.[6]

Atmospheric CO$_2$ Removal

Achieving negative emissions necessarily involves removing CO$_2$ from the atmosphere. Apart from bioenergy with CCS, various techniques have been proposed to accomplish this, although they are in early research stages.

Direct Air Capture

Although all techniques in this section capture CO$_2$, *direct air capture* usually refers to the use of chemical processes to extract CO$_2$ from the atmosphere, analogous to the way scrubbers capture CO$_2$ from the air inside spacecraft. Unlike bioenergy with CCS, these systems do not use large amounts of land, so they would not pose food security or deforestation risks.

Direct air capture systems need a lot of energy. To achieve negative carbon emissions, direct air capture systems must be powered by emission-free energy, such as wind, solar, or nuclear power, and that energy must not be taken from other users who would then rely on fossil energy instead. (That is, the emission-free energy used by a direct air capture system must be strictly additional to other emission-free energy uses.)

The other challenge facing direct air capture systems is cost. The estimated cost of a system that captures 1 million tons of CO$_2$ per year (roughly 0.02 percent of annual U.S. emissions) was $2.2 billion as of 2011. Over the plant's lifetime, the all-in cost is $600 per ton of CO$_2$, roughly eight times higher than the cost per ton to capture CO$_2$ from the flue gas of a coal power plant.[7] (Exhaust streams feature higher CO$_2$ concentrations, which makes the CO$_2$ easier to capture.)

Research can help improve the energy efficiency and lower the capital cost of direct air capture systems. As with other technologies to remove CO$_2$ from the atmosphere, carbon pricing can provide an economic incentive and the possibility of financial returns.

Enhanced Weathering

In nature, when certain types of minerals (such as olivine) are exposed to air and water, they undergo chemical reactions that extract CO$_2$ from the atmosphere and store it as a carbonate mineral.[8] These minerals make their way to the ocean, where organisms use the minerals to form shells and skeletons. When the organisms die, the material sinks into the deep ocean and eventually may be converted to limestone.[9]

Although this natural process is too slow to help reduce atmospheric CO_2 concentrations on human timescales, it may be possible to accelerate the natural process. For example, if large quantities of olivine and similar minerals were mined, finely ground (to increase their surface area), and spread on beaches or other land exposed to water and the atmosphere, the rate of CO_2 capture could be accelerated.[10]

Unfortunately, given current scientific understanding, the amount of olivine we would need to use would be very large, and the mining, transport, grinding, and spreading of the olivine would have to be done in a manner that releases few if any carbon emissions in order to achieve net sequestration. Additionally, for the sequestration to be sufficiently rapid, the olivine may have to be ground to particles with a mean diameter of less than 10 microns,[11] a microscopic size that is easily aerosolized and could be inhaled (as PM_{10}). More research would be needed to develop improved techniques before enhanced weathering could be considered a viable option for CO_2 removal on human timescales.

Ocean Fertilization

Phytoplankton are photosynthetic organisms in the ocean that extract CO_2 from seawater to build their bodies. When plankton die, they sink to the ocean floor, sequestering the CO_2 in their bodies.

Like other organisms, phytoplankton need a variety of nutrients to survive. In many parts of the ocean, iron is the limiting nutrient that constrains phytoplankton growth.[12] Therefore, it has been proposed that the ocean may be seeded with iron, encouraging phytoplankton growth, as a means of accelerating CO_2 sequestration.

There are a number of challenges with this approach. Many phytoplankton produce toxins, so encouraging their growth could lead to an increase in harmful algal blooms that threaten the health of marine ecosystems (and can harm or kill humans who eat contaminated seafood).[13] Also, when phytoplankton die, the bacteria that decompose them may deplete the oxygen in the water, leading to a "dead zone" that suffocates animal life.[14] Finally, algal growth in one area can inhibit algal growth in another area, and nutrients other than iron may become limiting nutrients in some places, so the effectiveness of iron seeding at increasing overall phytoplankton numbers has been questioned.[15]

More research could help determine whether ocean fertilization can be done safely and whether it offers significant CO_2 removal potential.

Biofuels and Synthetic Fuels for Transport

It may be possible to electrify many forms of transport, such as light-duty on-road vehicles. However, it can be difficult to electrify certain transport options, such as commercial aircraft, because of the requirements for fuel of high energy density. One option for these vehicle types is a carbon-neutral biofuel or other synthetic fuel.

Ethanol is a biofuel that is already widely used for transport, but ethanol derived from corn offers only 20 percent less greenhouse gas emissions than petroleum gasoline on a lifecycle basis,[16] and vehicles cannot run on 100 percent ethanol without special engine designs. To achieve zero emissions, a biofuel must be carbon neutral on a lifecycle basis, and it is more likely to be adopted if it is a drop-in replacement for gasoline (or diesel).

Ethanol made from cellulose, the inedible substance that forms the leaves and stalks of plants, can be made from agricultural residues rather than corn, lowering lifecycle greenhouse gas emissions. Various companies have experimented with obtaining biofuels from algae, although most of these businesses failed or pivoted to higher-value products, such as cosmetics or food additives, upon realizing the magnitude of the technical challenges involved.[17]

Another approach involves creating fuels directly from sunlight using a chemical or biological process.[18] This has the potential to avoid the inefficiency of using plants to convert sunlight into biomass, then converting that biomass to an energy-dense liquid fuel. These approaches are all in research stages and will need long-term, consistent government support to have a chance at commercialization.

Hydrogen

It is also possible to use hydrogen as a chemical fuel. Hydrogen has several advantages over carbon-neutral biofuels. First, the technology is more mature. We are able to produce hydrogen today with little or no greenhouse gas impact by using electricity from renewables to split water into hydrogen and oxygen. Also, hydrogen does not emit any pollutants when used for energy; the only byproduct is water vapor. Even if carbon neutral, biofuels may emit particulates and other pollutants harmful to human health when burned.

An important downside of hydrogen is that, to achieve sufficient energy density for use in a vehicle, the chemical must be stored at very high pressure or very low temperature. This necessitates a bulky and heavy storage system.

For example, the first ship to use hydrogen as a fuel (albeit not its primary fuel) was recently ordered. One of the technical challenges of this ship design is to store liquid hydrogen at $-253°C$.[19] As a result, hydrogen is more likely to be used in shipping or long-distance land-based transportation than in aircraft. The development of a large hydrogen distribution network and fueling stations would also be necessary.

Similarly, transporting hydrogen over large distances for use in equipment would require a new pipeline network. In some regions, such as the United States, it could be very difficult and expensive to construct the new pipeline infrastructure necessary to transport hydrogen for use across the economy.

The main thing governments can do to further the use of hydrogen is to promote the development of standards for hydrogen use in ships and other vehicles, including through international bodies, such as the International Maritime Organization. Government can also facilitate the buildout of hydrogen fueling and distribution infrastructure, if it is merited by sufficient demand.

Dietary and Behavioral Change

Some sources of greenhouse gas emissions, such as enteric fermentation in ruminants, may be difficult to tackle through technology this century. One option for lowering some types of emissions is for the government to use policies to shift human diet or behavior. For example, ending subsidies to crops used for animal feed (such as corn) or taxing beef and lamb may reduce demand for these goods. If demand is sufficiently reduced, a ban may become feasible. Although this may sound politically unlikely, bans of this sort already exist and enjoy broad public support; for example, government restrictions make it impossible to slaughter horses for meat in the United States.[20]

Behavioral change is not limited to targeting emission sources with few technological options; it may be helpful in a broad variety of circumstances. For example, zoning may be used to encourage walking and biking rather than driving private cars (discussed in more detail in Chapter 9).

Albedo (Reflectivity) Modification

Climate change is caused by the increase in heat-trapping gases in Earth's atmosphere. Radiant heat is infrared radiation, which is emitted by the surface of Earth after it absorbs sunlight and increases in temperature. If sunlight is reflected rather than absorbed, it may reenter space, bypassing the heat-trapping gases in the atmosphere. Therefore, one approach to tackle climate change is to increase the albedo (reflectivity) of Earth, so less sunlight is absorbed by the surface and turned into heat.

The main proposed mechanism of albedo modification is to inject tiny particles into the stratosphere, where they would act like a sunshade, scattering a fraction of the sunlight back into space.[21] This effect can occur in nature after a large volcanic eruption, which can release large quantities of particle-forming sulfur dioxide.[22] This work is still in the R&D stage and could be accelerated through government support.

Other mechanisms involve increasing the albedo of Earth's surface, such as by using light colors for rooftops and pavement. Albedo ranges from 0.0 (perfectly absorbing) to 1.0 (perfectly reflective). In urban areas, each 0.01 increase in albedo over a square meter of surface area results in a cooling effect equivalent to avoiding 7 kg of CO_2 emissions.[23] High-albedo surfaces can also mitigate the urban heat effect and reduce the need for air conditioning in warm climates, saving energy and potentially reducing emissions. Government can promote the use of high-albedo materials through building codes and through direct procurement for public roadway and sidewalk materials.

Policies for Adapting to a Warmer World

When planning for a post-2050 future, it is wise to make investments in technologies and measures that will help humanity adapt to climate change. Adaptation can be categorized as structural or physical (e.g., building seawalls or creating new crop or animal varieties), social (e.g., evacuation planning or preparing for migration flows), or institutional (e.g., insurance ownership requirements or urban planning for climate change).[24] Many of these measures require long lead times, and efforts to tackle them should begin now. For example, it may be desirable to genetically engineer new varieties of crops that are resistant to drought, heat waves, or different sorts of pests yet are safe for the environment. Work to develop suitable crops may take many years. Governments should provide R&D support for adaptation technologies today.

Governments necessarily take the lead on resilient urban planning, disaster planning, construction of protective infrastructure (e.g., seawalls and early warning systems), and other institutional measures. National governments may require the participation of local governments, particularly for measures best undertaken at the local level, such as by stipulating that cities must take climate change into account in their urban plans. Insurance against natural hazards can be required by government, and premiums may be structured in a way to encourage people to reduce their exposure to damage from extreme weather events.

Governments also must plan for both internal and cross-border migration. The United Nations High Commissioner for Refugees has estimated that "up to 250 million people may be displaced by the middle of this century as a result of extreme weather conditions, dwindling water reserves . . . a degradation of agricultural land . . . [and] to escape fighting over meagre resources."[25] This is roughly 50 times the number of refugees from the Syrian civil war, or 3.2 percent of today's global population. (Lest this figure sound too high to be credible, consider that the United Nations already tracks 64 million refugees, asylum seekers, and internally displaced people, many of whom were forced to relocate by problems caused or exacerbated by climate change.[26]) Working to ensure that adequate infrastructure, housing, and job opportunities are available, and that migrants can be successfully integrated into new societies, will be critical to preserving political stability. Without proper preparations, the reaction to refugee flows could compound the physical damages caused by climate change.[27]

Conclusion

Enacting strong policy to mitigate emissions as soon as possible offers the most cost-effective opportunities to limit damage to human societies. However, human needs in a post-2050 world will be different from human needs in 2020 or 2030, and the technology necessary to satisfy those needs does not exist at present. Therefore, even as we act to cut emissions today, it is crucial to begin work on strategies with long lead times to obtain the technologies we will need in the latter part of the century and beyond. In particular, research into CCS technology, various ways to remove CO_2 from the atmosphere, and certain high-importance adaptation measures (e.g., genetically engineered crops, to avoid famine) should be robustly supported by present-day government policy and investment. This will help to ensure we have a full portfolio of options with which to tackle future challenges as they arise.

Conclusion

Immediate reductions in greenhouse gas emissions are needed to avoid the worst impacts of climate change. Failing to rapidly reduce emissions could result in significant damage: loss of coastal lands to sea level rise, threatening more than a billion people; mass refugee migrations; famines; a wave of extinctions; and other impacts that will take an economic, ecological, and human toll.

Fortunately, new technologies continue to show that a low-carbon future is within reach and perhaps as cheap as or even cheaper than a high-carbon one. Falling costs of renewable energy technologies—such as solar panels, onshore and offshore wind turbines, super-efficient LED light bulbs, and electric vehicles—mean that the cost of a low-carbon economy keeps falling. In fact, in many regions it is already cheaper to build and operate new zero-carbon electricity technologies than to continue to operate old polluting ones—even ones that are already built![1] And in nearly every region, clean energy deployment continues to outpace even the most aggressive forecasts. For example, consider the fact that Shenzhen, a Chinese city of 12 million people, just finished converting all 16,359 of its buses to run on electricity.[2] Just 5 years ago, this was unthinkable.

Although the technology exists today and is falling in cost, significantly reducing global greenhouse gas emissions is a Herculean task, one that will not simply happen on its own even with cheap clean technology. To achieve the reductions needed to keep warming below two degrees and avoid the worst impact of climate change, the highest-emitting countries must rapidly adopt the most effective energy policies and design them well.

Most greenhouse gas emissions—nearly 75 percent—come from just 20 countries. The source of these emissions is predominantly energy use (e.g., power plants, vehicles, and buildings) and industrial processes (e.g., cement or iron and steel manufacturing). The focus should therefore be on reducing emissions from energy use and industrial processes.

Four types of energy policies can be used to address these emission sources: performance standards, economic signals, support for research and development (R&D), and enabling policies. Designed well, these policy types interact with and reinforce one another, strengthening a portfolio of policies to deliver deeper and more cost-effective emission reductions.

Performance standards set minimum requirements for energy efficiency, renewable energy uptake, or product performance. Examples include vehicle fuel economy standards, energy-efficient building codes, renewable portfolio standards, and emission limits for power plants.

Economic signals are policies designed to accelerate the adoption of clean energy technologies, ensure that positive and negative social impacts (i.e., externalities) are incorporated into product costs, or otherwise use the market as a tool for efficiently achieving emission reductions. Examples include carbon taxes and subsidies for clean energy production or efficiency upgrades.

Government support for R&D can accelerate innovation. New technology spurs economic development and reduces reliance on expensive and volatile fossil power sources. Government support can come in the form of funding for basic research (on technologies far from commercialization) intended to benefit many new industries. However, one of the most powerful ways government can support R&D involves creating an environment where private sector R&D can thrive. Examples include sharing technical expertise and facilities (such as national laboratories); adopting appropriate intellectual property protections; promoting robust science, technology, engineering, and mathematics (STEM) education in public schools and universities; and structuring immigration laws so companies are not prevented from hiring foreign STEM talent.

Enabling policies enhance the functionality of the other policies, often through direct government expenditures, information transparency, or reduction of barriers to better choices. For example, a policy requiring clear energy use labels on products enables consumers to make smarter decisions, and good urban design gives people transit options other than driving their cars, enabling them to respond to well-designed economic signals.

For energy and climate policy to be effective, a suite of policies is needed; there is no silver bullet in this business. To design an optimal suite of policies, a policymaker should consider policies from each of these categories. Together, they create a powerful symbiosis that can drive deeper carbon emission reductions than policies in isolation while increasing overall cost-effectiveness of emission reduction.

Deciding which policies to implement is often controversial. Which will deliver substantial emission reductions? What will the costs be? How do certain policies interact with others?

The first step in evaluating which policies to prioritize is to assess the structure of the economy and emissions. Knowledge about how many cars, buildings, and power plants there are, how much energy they use, how their use is expected to grow over time, and so on can highlight which areas of the economy must be a focus for emission abatement.

The next step is to quantitatively evaluate the potential for policies to reduce emissions. Fortunately, advances in modeling, especially when combined with decades of experience in designing and implementing energy policy, now allow us to quantify which policies are most effective at reducing carbon emissions and their relative costs. Marginal abatement cost curves and policy cost curves provide policymakers with detailed information on which technologies and policies have the most potential to reduce emissions, which technologies and policies can reduce emissions most cost-effectively, and how policies interact with one another, either undercutting or amplifying emission reductions. Not every country needs to conduct all these analyses. Some regions will be quite similar to others that have already conducted some or all of these assessments, in which case similar conclusions will apply.

This modeling has also elucidated the fact that just a small set of policies has the potential to significantly reduce emissions from energy and industrial processes. These policies, implemented stringently in the highest-emitting countries, can put us on the path to meeting the two-degree target.

In the power sector, renewable portfolio standards and feed-in tariffs can reduce emissions by increasing the share of fossil-free power generation. Designed well, they can minimize the costs of transitioning to a low-carbon power system. Complementary policies, such as support for transmission lines, smart utility policy structures, and efficiency resource standards, are important as well.

Strong standards and incentives to improve energy use in industry can significantly reduce emissions. The industry sector also produces significant process emissions, which include emissions of CO_2 and other non-CO_2 gases generated in industrial processes. Policymakers can require control of these emissions and offer incentives and other forms of assistance—technical, financial, or other—to encourage reductions.

In the building sector, building codes and appliance standards are the best tools for reducing emissions. Economic signals do not work well in the building sector, because of well-known market failures that discourage building owners and renters from taking cost-saving measures. Instead, building codes and standards tend to save money, as decreased energy use outweighs any increase in costs over time.

Fuel economy standards, vehicle feebates, electric vehicle incentives, and smart urban planning can all reduce emissions significantly in the transportation sector by increasing the fuel efficiency of vehicles, decreasing their emissions, and offering alternative transportation options.

Carbon pricing is another strong tool for reducing emissions, encouraging emission-reducing behavior across the economy and pushing investments to lower-carbon options.

Finally, support for R&D helps reduce the costs of all of these policies while providing opportunities for new low-carbon technologies to hit the market.

Of course, these policies must be designed well if they are to achieve real-world reductions. Decades of experience with both good and bad policy design has illuminated the characteristics that separate good from bad policy. For example, without built-in mechanisms for continuous improvement, policies tend to stagnate and become obsolete. And without a sufficiently long time horizon, businesses cannot invest in the technology or R&D needed to produce better equipment. The handful of policy design principles discussed in Chapter 2 can ensure that future climate and energy policy maximizes greenhouse gas emission reductions and economic efficiency. These principles can drive effective, investment-grade policy.

We must also have an eye toward what is necessary over the long term. Policymakers should consider the technologies needed after 2050 to drive deep decarbonization, including achieving negative emissions as is required by nearly all two-degree emission scenarios. To this end, policymakers must support R&D initiatives to bring these technologies to market (while pursuing aggressive energy policies to drive cleaner and more efficient power plants, cars, factories, and buildings in the meantime).

* * * *

Climate change requires action as soon as possible to limit emissions and avoid their worst impacts. Governments, businesses, and organizations around the world committed to reducing emissions under the Paris Agreement, laying the foundation for emission cuts that put the world on a trajectory for a lower-carbon future. The key now is in turning these pledges into reality—with laser-focused, well-designed policy.

This is a large task. Not only is it possible—*we know how to do it*. We have the technology today to rapidly move to a clean energy system. And the price of that future, *without counting environmental benefits*, is about the same as that of a carbon-intensive future. So the challenge is not technical, nor even economic, but rather is a matter of enacting the right policies and ensuring they are properly designed and enforced. We hope this book helps policymakers and others take the next step of turning these pledges into action to achieve the deep emission reductions that we so desperately need.

The Energy Policy Simulator

The Energy Policy Simulator (EPS) is a system dynamics computer model that estimates the effects of various policies[1] on emissions, financial metrics, electricity system structure, and other outputs. The EPS model is designed to represent different countries by incorporating input data specific to the country in question. This Appendix discusses the purpose, structure, and function of the EPS. More detail on the technical workings of the EPS is available in the EPS online documentation.[2]

The EPS was developed by Energy Innovation, LLC with help from the Massachusetts Institute of Technology and Stanford University. The model has been peer reviewed by people associated with Argonne National Laboratory, the National Renewable Energy Laboratory, Lawrence Berkeley National Laboratory, Stanford University, China's National Center for Climate Change Strategy and International Cooperation, China's Energy Research Institute, and Climate Interactive.

The EPS aims to help policymakers evaluate a wide array of climate-related policies. The tool allows users to explore unlimited policy combinations and to adjust policy levers to any setting, allowing them to create their own policy scenarios. It simulates the years 2017–2050, using annual time steps, and offers hundreds of outputs. Some of the most important are emissions of 12 different pollutants; cash flow (costs and savings) for government, industry, and consumers;[3] capacity and generation of electricity by different types of power plants; land use changes and associated emissions or sequestration; and premature deaths avoided by reductions in particulate emissions. These outputs could help policymakers anticipate long-term impacts and costs of implementing new policies. Many of the policies included in the EPS have not yet been explored in many countries, helping to present new options to policymakers. The tool may not only help inform a roadmap for policymakers to implement climate goals (e.g., from the Paris Agreement), but it may also show how policymakers could set new goals and increase their country's ambitions.

The model is free and open-source. It can be used via an interactive web interface at https://energypolicy.solutions and can be downloaded from the same site.

Why Use a Computer Model to Assist in Policy Selection?

Before considering the structure and uses of the EPS, it is worthwhile to ask, "Why do we use a computer model at all?"

A policymaker seeking to reduce emissions faces a dizzying array of policy options that might advance this goal. Policies may be specific to one sector or type of technology (e.g., light-duty vehicle fuel economy standards) or might be economy-wide (e.g., a carbon tax). Sometimes a market-driven approach, a direct regulatory approach, or a combination of the two can be used to advance the same goal. For instance, to improve the efficiency of home appliances, a government might offer rebates to buyers of efficient models, mandate that the appliance manufacturers meet specific energy efficiency standards, or both. To navigate this field of options, policymakers need an objective, quantitative mechanism to determine which policies will meet their goals and at what cost or savings.

Many studies have examined particular energy policies in isolation. However, it is of greater value to policymakers to understand the effects of a package of different policies because the policies may interact. This interaction can produce results different from the sum of the effects of the individual policies.

Thanks to the strength of computer models at simulating complex systems, a customized computer model is a crucial tool to help policymakers evaluate a wide array of policies. To understand how to hit emission targets that involve emission reductions from every sector, a satisfactory model must be able to represent the entire economy and energy system with an appropriate level of disaggregation, be easy to adapt to represent new countries, be capable of representing a wide array of relevant policy options, and offer results that include a variety of policy-relevant outputs. Additionally, the model must capture the interactions of policies and other forces in a system whose parameters change dramatically over the course of the model run, as countries continue to grow and develop.

About System Dynamics Modeling

A variety of approaches exist for representing the economy and the energy system in a computer simulation. The EPS is based on a theoretical framework called system dynamics. As the name suggests, this approach views the processes of energy use and the economy as an open, ever-changing, nonequilibrium system. This may be contrasted with approaches such as computable general equilibrium models, which regard the economy as an equilibrium system subject to exogenous shocks, or disaggregated technology-based models, which focus on the potential efficiency gains or emission reductions that could be achieved by upgrading specific types of equipment.[4]

System dynamics models often include "stocks," or variables whose value is remembered from modeled year to modeled year, and which are affected by "flows" into and out of these variables. For example, a "stock" might be the total installed capacity of coal power plants, which can only grow or shrink gradually, because of construction of coal plants (an inflow) and retirement of old plants (an outflow). In contrast, the amount of energy generated by coal plants in a given year is calculated afresh every year and is therefore not a "stock" variable.

System dynamics models often use the output of the previous time step's calculations as input for the next time step. The EPS follows this convention, with stocks such as the electricity generation fleet, the types and efficiencies of building components, and so on remembered from one year to the next. Therefore, an efficiency improvement in an early year will result in fuel savings in all subsequent years until the improved vehicle, building component, or other investment is retired from service.

The industry sector is handled differently. Because the available input data come in the form of business-as-usual levels of fuel use and potential reductions in fuel use and process-related emissions by policy, we gradually implement these reductions (with corresponding implementation costs) rather than recursively tracking fleet-wide efficiency. Because of the diverse forms input data can take in the sectors we model, one approach rarely works for all sectors. Accordingly, the EPS attempts to use whichever approach makes the most sense in the context of a specific sector.[5]

Structure of the EPS

The EPS structure can be envisioned along two dimensions: the visible structure that pertains to the equations that define relationships between variables (viewable as a flowchart) and a behind-the-scenes structure that consists of arrays (matrices) and their elements, which contain data and are acted on by the equations. For example, the transportation sector's visible structure consists of policies (such as a fuel economy standard), input data (such as the kilometers traveled by a passenger or a ton of freight, or the elasticity of travel demand with respect to cost), and calculated values (such as the quantity of fuel used by the vehicle fleet). The arrays in the transportation sector consist of vehicle categories (light-duty vehicles [LDVs], heavy-duty vehicles [HDVs], aircraft, rail, ships, and motorbikes), cargo types (passengers or freight), fuel types (e.g., petroleum gasoline, petroleum diesel, electricity), and vehicle engine types (e.g., gasoline engine, diesel engine, electric engine). The model generally performs a separate set of calculations, based on each set of input data, for every combination of array elements. For example, the model will calculate different fuel economies for passenger HDVs, freight HDVs, passenger aircraft, freight aircraft, and so forth.

The model has five main sectors (industry and agriculture, buildings, transportation, electricity, and land use), plus a few supporting modules handling other functions, as depicted in Figure I-1.[6]

The model's calculation logic begins with the fuels section, where basic properties of all fuels are set and policies that affect the price of fuels are applied. Information about the fuels is used in the three demand sectors: transportation, buildings, and industry and agriculture. These sectors calculate their own emissions from direct fuel use (e.g., fossil fuels burned in vehicles, buildings, and industrial facilities). These sectors also specify a quantity of electricity or heat (energy carriers supplied by other parts of the model) consumed each year. The electricity sector and the district heat module consume fuel to supply the energy needs of the three demand sectors, accounting for transmission and distribution losses. The fifth sector, land use, does not consume fuel or electricity.

All five sectors and the district heat module produce emissions of each pollutant, which are summed, represented by the pollutants box at the bottom of Figure I-1. The same is true for cash flow impacts, which are calculated separately for particular actors (government, industry, consumers, and several specific industries). Calculation of changes in spending (e.g., on capital equipment, fuel, and operations and maintenance), as well as monetized social benefits from avoided public health impacts and climate damages, are also carried out at this stage.

Two model components affect the operation of various sectors. A set of research and development levers allows the user to specify improvements in fuel economy and decreases in capital cost for technologies in each of the four sectors and in the carbon capture and sequestration (CCS) module. The CCS module alters the industry and electricity sectors by reducing their CO_2 emissions (representing sequestration), increasing their fuel usage (to power the energy-intensive CCS process), and affecting their cash flows.

Input Data Sources

The model has significant input data requirements, necessitating the use of a variety of data sources. All regional EPS models are adapted from the international, open-source release of Energy Innovation's Energy Policy Simulator. To adapt the model to a new region, input data are sourced via the following approaches, in order of priority:

- National data located in published sources, produced as outputs from other models or provided by national governments.
- When national data are not available, input data from other countries[7] are scaled to represent the country being built. Scaling factors differ by variable

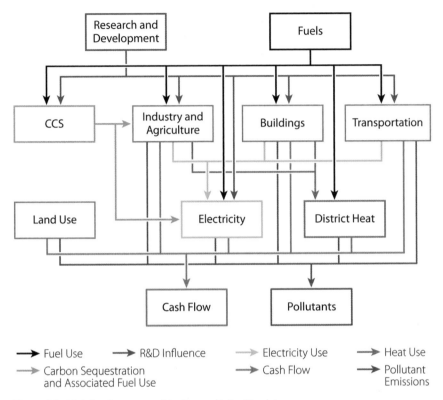

Figure I-1. High-level structure of the Energy Policy Simulator.

and are selected based on which scaling factor most closely correlates with the variable in question. For example, a variable pertaining to economic output or production might be scaled by GDP, whereas a variable related to wastewater treatment might be scaled by population.

- When national data are unavailable and scaling other countries' data is irrelevant or inappropriate, other countries' input data are used unchanged. Scaling data is irrelevant when the data are not actually country-specific (e.g., the global warming potentials of various gases). It may also be inappropriate when, for example, the expected lifetime of a building component (such as an air conditioner) in another country may be the best available estimate of the lifetime of that same type of building component in the country being modeled. Scaling the lifetime of a foreign air conditioner by any available factor (e.g., population, GDP) would be nonsensical.

In general, the model uses only publicly available data and sources that do not cost money to access. All of the data used in the model are documented and sourced meticulously. Many variables have more than one data source, so full source information can sometimes be extensive. Full source information is available in each variable's associated spreadsheet file, which can be downloaded as part of the EPS package.

Methodology for Quantitative Policy Assessment

This book estimates the potential of individual policies to meet a future emission and warming scenario. These estimates are not meant to represent the exact abatement potential achievable in each of the regions modeled. However, they are supposed to indicate the relative magnitude of different policy options and to demonstrate the value in having a broad suite of policies with strong early action and continuous improvement.

Reference Scenario Emissions

Ideally, forecasting emissions and potential reductions would rely on emission projections from each country through 2050. However, most countries do not have projections to 2050. Even when countries do have projections to 2050, they often omit certain sectors (e.g., land use or industrial processes) or use different global warming potential values. Given the paucity of 2050 emission projections and the heterogeneity of the projections that do exist, a country-level analysis was not possible.

Instead, we relied on regional emission modeling completed under the Low Climate Impact Scenarios and the Implications of Required Tight Emission Control Strategies (LIMITS) modeling, initiated by the European Union and published in 2013.[1] LIMITS modeling has been used to assess previous climate negotiation scenarios relying on the same models used extensively in the Intergovernmental Panel on Climate Change (IPCC) Fifth Assessment Report (AR5). Of the major climate modeling exercises, it has the finest level of detail, with 10 superregions (whereas most other modeling efforts evaluate only 5 regions) and the most up-to-date forecasts at the time of this writing.

The LIMITS modeling exercise included several different model teams from around the world. Based on the availability of sector-specific results and alignment with other modeling, we chose to use the Pacific Northwest National Laboratory

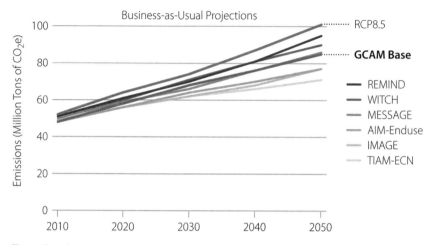

Figure II-1. Business-as-usual emission projections. (Analysis done using data with permission from the International Institute for Applied Systems Analysis [IIASA]. Data sources: Clarke et al., 2014. Data downloaded from the IPCC-IAMC database hosted at IIASA, https://secure.iiasa .ac.at/web-apps/ene/AR5DB; Data source: Tavoni et al., 2013. Data downloaded from the LIMITS Scenario database hosted at IIASA, https://tntcat.iiasa.ac.at/LIMITSPUBLICDB/dsd?Action=html page&page=about.)

and the Joint Global Change Research Institute's model results using the Global Change Assessment Model (GCAM).

For our reference scenario, we used the "Base" scenario, which assumes no near-term or global action on emissions through 2100. Figure II-1 shows how the Base scenario for the GCAM compares with the other models used for LIMITS and the Representative Concentration Pathway (RCP) 8.5 Scenario, the business-as-usual scenario in AR5.

Under the Base scenario, global greenhouse emissions grow from 48.0 Gt CO_2e in 2010 to 84.7 Gt CO_2e in 2050, using global warming potential values from the IPCC Fourth Assessment Report (AR4) and covering CO_2, methane, nitrous oxide, and F-gases. We used values from AR4 because emission values for F-gases are available only in aggregate CO_2e and therefore cannot be adjusted to values used in the AR5. The GCAM emission trajectory is roughly a mid-case scenario across the different models.

2050 Target Emission Level

The LIMITS study mapped two sets of climate policy targets: a 450-ppm scenario with radiative forcing of 2.8 W/m² , giving a greater than 70 percent chance of staying under two degrees of warming; and a 500-ppm scenario with radiative forcing of 3.2 W/m², giving a 50 percent chance of staying under two degrees of warming.[2]

The LIMITS modeling evaluated different scenarios for achieving those targets. These include a benchmark scenario, a Reference Policy scenario ("Ref Pol"), and a Strong Policy ("StrPol") scenario, in addition to variations on each of these.

The Ref Pol scenario assumes a weak near-term target with fragmented action until 2020. The StrPol scenario assumes a stringent near-term target with fragmented action until 2020.

We used the Ref Pol-500 scenario, which combines the target and near-term action assumptions of the Ref Pol scenario with the emission target of the 500 scenario, in assessing necessary emission reductions through 2050. The Ref Pol-500 scenario is a good policy scenario because it assumes some limited action through 2020, reflective of the current state of affairs on global climate policy, with significant reductions after 2020 to achieve the 500-ppm target.

For context, the GCAM Ref Pol-500 scenario is one of the more aggressive emission reduction scenarios between 2020 and 2050. It falls roughly in between the RCP2.6 and RCP4.5 scenarios from AR5. Figure II-2 shows how the GCAM RefPol-500 scenario compares with these other low-emission scenarios.[3]

Using the Ref Pol-500 scenario as our policy target, we estimated necessary reductions by calculating the cumulative emission reductions between 2010 and 2050 between the Base scenario and the Ref Pol-500 scenario. Although the LIMITS modeling runs to 2100, we limited our assessment to 2050 in order to use the Energy Policy Simulator (EPS) for the analysis, discussed in Appendix I.

Next, we estimated real business-as-usual emissions between 2010 and 2020 using observed data on emissions between 2010 and 2014[3] and extrapolating this trend to 2020.

Accounting for the difference in observed/forecasted business-as-usual emissions between 2010 and 2020 and the GCAM emission projection over this period, we then calculated the total emission reductions needed between 2020 and 2050 from the LIMITS modeling. This value, a 43 percent reduction in cumulative emissions, served as our emission target.

Mapping onto Reference Countries

The EPS, discussed in detail in Appendix I, served as the primarily tool for calculating the potential of policies to reduce emissions. To use the EPS, we first mapped each sector in each of the 10 superregions from the LIMITS modeling onto the sector of an existing EPS model. At the time of the analysis, we had models for the United States, China, Indonesia, Poland, and Mexico. We could not use the

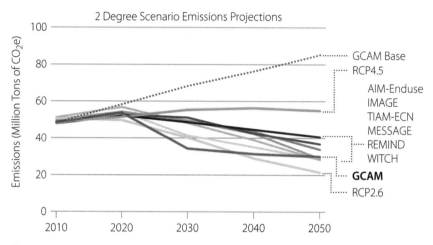

Figure II-2. 2°C scenario emission projections. (Analysis done using data with permission from the International Institute for Applied Systems Analysis [IIASA]. Data sources: Clarke et al., 2014. Data downloaded from the IPCC-IAMC database hosted at IIASA, https://secure.iiasa.ac.at/web-apps/ene/AR5DB; Data source: Tavoni et al., 2013. Data downloaded from the LIMITS Scenario database hosted at IIASA, https://tntcat.iiasa.ac.at/LIMITSPUBLICDB/dsd?Action=htmlpage&page=about.)

Mexico model because it evaluates emissions only to 2030, whereas the other models evaluate emissions to 2050.

Each of the sectors in the 10 superregions was mapped onto the same sector in either the United States, China, Indonesia, or Poland model, based on how closely the rate of emission growth (or decline) in that sector matched the same sector in one of the four EPS models. In some instances, expert judgment was used to map a sector onto the most similar country model. We also mapped each region onto a country as a whole to assess cross-sector policies, such as carbon pricing.

Representative Country Abatement Potential

Policy scenarios were then developed for each of the reference country EPS models, using similar policy settings. In some instances unique policy settings were used that reflect the unique abatement options available to different country types. For example, a ban on new coal power plants was used in Indonesia, a region that has a large forecasted increase in electricity demand. However, this policy was not used in China, the United States, or Poland, which all have flat electricity demand or are already oversupplied. A full list of the policies and settings used is included in Table II-1. Note that the policy settings included are 2030 values. After 2030, where applicable, policy settings increase at 1 percent per year through 2050. In some cases the policy settings may seem arbitrary, but they were set to achieve uniformity (e.g., fuel efficiency standards for light-duty vehicles were designed to achieve 70 miles per gallon in 2050 across all models).

Potential Reductions by Policy

We then calculated the percentage reduction in sectoral emissions in each of the EPS models for each policy. For example, we calculated the annual percentage reduction in emissions in the power sector from a renewable portfolio standard of 40 percent in the "U.S. EPS model" to avoid confusion. We did this assessment for every policy enabled. For cross-sector policies, such as a carbon tax, we looked at the reduction in economy-wide emissions after accounting for reductions from the sector-specific policies.

We assumed that each of the superregions could achieve the same percentage reduction in sectoral emissions as the region it was mapped onto. For example, if the renewable portfolio standard of 40 percent resulted in power sector emission reductions of 30 percent in 2050 in the U.S. EPS, then it was assumed each region that was mapped to the U.S. power sector could achieve a 30 percent reduction in that region's power sector emissions through 2050. A similar approach was applied for cross-sector policies.

We then summed the emission reductions across regions by policy type through 2050 to develop cumulative emission reduction estimates through 2050.

In aggregate, we found that a small set of policies, set stringently, could achieve large enough cumulative reductions through 2050 to meet the reductions achieved under the Ref Pol-500 scenario.

An important note to this analysis is that our method looks at the ability of new policies or strengthened existing policies to drive further emission reductions. The analysis does not attribute the emission reductions from existing policies to the policy category being assessed. For example, China, Poland, and the United States all have strong vehicle fuel economy standards on the books. Our modeling evaluates the potential for additional emission reductions on top of those that will be achieved with existing policy. Contextualizing the results within the existing policy framework is critical. Policies that have already been enacted stringently are likely to show smaller emission reductions potential than some might expect, because this analysis looks only at the incremental reductions possible.

Table II-1
Energy Policy Simulator Policy Settings

Policy Group	Policy Name	Implementation	Policy Setting in 2030 (Grows at 1% / Year Thereafter)			
			China	Indonesia	Poland	United States
Building codes and appliance standards	Appliance Standards Efficiency Improvement	% Improvement in new equipment efficiency	35%	35%	35%	35%
	Building Code Efficiency Improvement: Cooling	% Improvement in new equipment efficiency	35%	35%	35%	35%
	Building Code Efficiency Improvement: Envelope	% Improvement in new equipment efficiency	30%	30%	30%	30%
	Building Code Efficiency Improvement: Heating	% Improvement in new equipment efficiency	30%	30%	30%	30%
	Building Code Efficiency Improvement: Lighting	% Improvement in new equipment efficiency	35%	35%	35%	35%
	Building Code Efficiency Improvement: Other Components	% Improvement in new equipment efficiency	10%	10%	10%	10%
	Building Component Electrification	Share of newly sold equipment	50%	50%	50%	50%
	Building Contractor Education and Training	% Potential energy use reduced	100%	100%	100%	100%
	Building Retrofitting	% Additional stock retrofitted per year	2%	2%	2%	2%
	Improve Appliance Labeling	% Potential energy use reduced	100%	100%	100%	100%
	Rebates for Efficient Equipment	% Potential energy use reduced	100%	100%	100%	100%
Carbon pricing	Carbon Pricing	$/Metric ton CO_2e	$60	$60	$60	$60
Complementary power sector policies	Additional Demand Response	% Additional potential	50%	50%	50%	50%
	Ban New Coal Power Plants	n/a		Yes		
	Early Retirement of Coal Power Plants	Gigawatts per year	2.5	1		
	Least-Cost Electricity Dispatch	n/a	Yes	Part of BAU	Part of BAU	Part of BAU
	Transmission Growth	% Existing transmission capacity added	75%	75%	75%	75%
Feebate	Feebate	% Global best practice fee	75%	75%	75%	75%

Policy Group	Policy Name	Implementation	China	Indonesia	Poland	United States
						Policy Setting in 2030 (Grows at 1% / Year Thereafter)
Fuel economy standards	Fuel Economy Standards for HDVs	% Additional improvement in new vehicle efficiency	82%	82%	82%	82%
	Fuel Economy Standards for LDVs	% Additional improvement in new vehicle efficiency	15%	160%	18%	75%
	Fuel Economy Standards for Motorbikes	% Additional improvement in new vehicle efficiency	40%	40%	40%	40%
Industry energy efficiency	Cogeneration and Waste Heat Recovery	% Potential emission reductions	75%	75%	75%	75%
	Conversion of Non-CHP Heat to CHP	% Potential emission reductions	75%	75%	75%	75%
	Early Retirement of Industrial Plants	% Potential emission reductions	75%	75%	75%	75%
	Improved Industry Worker Training	% Potential emission reductions	75%	75%	75%	75%
	Industrial Energy Efficiency Standards	% Reduction in energy use	15%	15%	15%	15%
	Industrial Fuel Switching from Coal to Natural Gas	% Industry coal use switched	25%	25%	25%	25%
	Industrial Fuel Switching from Natural Gas to Electricity	% Industry natural gas use switched	25%	25%	25%	25%
Industry process emissions policies	Cement Clinker Substitution	% Potential emission reductions	75%	75%	75%	75%
	Cropland Management	% Potential emission reductions	75%	75%	75%	75%
	Livestock Measures	% Potential emission reductions	75%	75%	75%	75%
	Methane Capture	% Potential emission reductions	75%	75%	75%	75%
	Methane Destruction	% Potential emission reductions	75%	75%	75%	75%
	Reduced F-Gases	% Potential emission reductions	90%	90%	90%	90%
	Rice Cultivation Measures	% Potential emission reductions	75%	75%	75%	75%
Renewable portfolio standard	Distributed Solar Carve-Out	Share of building electricity	15%	15%	15%	15%
	Renewable Portfolio Standard	% of Electricity generation	50%	50%	50%	50%
Urban mobility	Transportation Demand Management	% of Potential mode shifting	75%	75%	75%	75%
Vehicle electrification	Electrification of Motorbikes	% New vehicle sales	50%	50%	50%	50%
	Electrification of Passenger HDVs	% New vehicle sales	50%	50%	50%	50%
	Electrification of Passenger LDVs	% New vehicle sales	50%	50%	50%	50%

Note: BAU = business as usual; CHP = combined heat and power; HDV = heavy-duty vehicle; LDV = light-duty vehicle; n/a = not applicable.

Notes

Introduction

1. For more information, see Appendix II.

2. H.A. Biswas, T. Rahman, and N. Haque, "Modeling the Potential Impacts of Global Climate Change in Bangladesh: An Optimal Control Approach," *Journal of Fundamental and Applied Sciences* 8, no. 1 (2016): 1–19.

3. Umesh Adhikari, A. Pouyan Nejadhashemi, and Sean A. Woznicki, "Climate Change and Eastern Africa: A Review of Impact on Major Crops," *Food and Energy Security* 4, no. 2 (July 2015): 110–32, https://doi.org/10.1002/fes3.61.

4. "The Economic Risks of Climate Change in the United States" (Risky Business Project, June 2014), https://riskybusiness.org/site/assets/uploads/2015/09/RiskyBusiness_Report _WEB_09_08_14.pdf.

5. "Causes of Climate Change" (U.S. EPA, 2016), https://19january2017snapshot.epa.gov /climate-change-science/causes-climate-change_.html.

6. Nicholas Stern, "Stern Review: The Economics of Climate Change" (HM Treasury, 2006), http://mudancasclimaticas.cptec.inpe.br/~rmclima/pdfs/destaques/sternreview _report_complete.pdf.

7. Mike Munsell, "Solar PV Prices Will Fall below $1.00 per Watt by 2020" (Greentech Media, June 1, 2016), https://www.greentechmedia.com/articles/read/solar-pv-prices-to -fall-below-1-00-per-watt-by-2020.

8. Ryan Wiser, et al., "Forecasting Wind Energy Costs and Cost Drivers: The Views of the World's Leading Experts" (Lawrence Berkeley National Laboratory, 2016), https://emp.lbl .gov/publications/forecasting-wind-energy-costs-and/.

9. Lazard, "Lazard's Levelized Cost of Storage Analysis," Version 1.0 (2015), https://www .lazard.com/media/2391/lazards-levelized-cost-of-storage-analysis-10.pdf.

10. "Large Scale Battery Storage" (EPIS, April 5, 2016), http://epis.com/powermarket insights/index.php/2016/04/05/large-scale-battery-storage/.

11. International Energy Agency, "Lighting," accessed December 5, 2017, http://www.iea .org/topics/energyefficiency/lighting/.

12. "Energy Efficiency Market Report 2015" (International Energy Agency, 2015), http:// www.iea.org/publications/freepublications/publication/MediumTermEnergyefficiency MarketReport2015.pdf.

13. "Policy Solutions," Energy Innovation, accessed January 10, 2018, https://us.energy policy.solutions/.

14. "Clean Energy Solutions Center," accessed December 5, 2017, https://cleanenergyso lutions.org/.

15. "Paris Contributions Map" (CAIT Climate Data Explorer), accessed December 5, 2017, http://cait.wri.org/indc/.

16. "Companies Taking Action" (Science Based Targets), accessed December 5, 2017, http://sciencebasedtargets.org/companies-taking-action/.

17. "The World's Most Influential Companies, Committed to 100% Renewable Power" (RE100), accessed December 5, 2017, http://there100.org/home.

18. "Climate Action Tracker: Global Emissions Time Series," http://climateactiontracker.org.

19. "Divestment Commitments" (Fossil Free), accessed December 5, 2017, https://go fossilfree.org/divestment/commitments/.

Chapter 1

1. "Climate Change 2014 Synthesis Report: Summary for Policymakers" (Intergovernmental Panel on Climate Change, 2014), Table SPM1, https://www.ipcc.ch/pdf/assessment -report/ar5/syr/AR5_SYR_FINAL_SPM.pdf.

2. James H. Butler and Stephen A. Montzka, "The NOAA Annual Greenhouse Gas Index (AGGI)" (U.S. Department of Commerce/National Oceanic & Atmospheric Administration, 2017), Table 2, https://www.esrl.noaa.gov/gmd/aggi/aggi.html.

3. Elmar Kriegler et al., "What Does the 2°C Target Imply for a Global Climate Agreement in 2020? The LIMITS Study on Durban Platform Scenarios," *Climate Change Economics* 4, no. 4 (November 1, 2013): 1340008, https://doi.org/10.1142/S2010007813400083.

4. "About LIMITS" (Science for Global Insight, n.d.), https://tntcat.iiasa.ac.at/LIMITS PUBLICDB/dsd?Action=htmlpage&page=about.

5. "Paris Contributions Map," World Resources Institute, CAIT Climate Data Explorer, INDC Dashboard, http://cait.wri.org/indc/.

6. A rating of "Inadequate" means "if all governments put forward inadequate positions warming likely to exceed 3–4°C." For more information, see "Rating Countries," Climate Action Tracker, accessed December 13, 2017, http://climateactiontracker.org/countries.html.

7. A rating of "Medium" means "not consistent with limiting warming below 2°C as it would require many other countries to make a comparably greater effort and much deeper reductions." For more information, see "Rating Countries," Climate Action Tracker, accessed December 13, 2017, http://climateactiontracker.org/countries.html.

8. The energy category includes electricity and heat production, manufacturing and construction, transportation, other fuel combustion, and fugitive emissions.

9. For mor information, see CAIT Climate Data Explorer, 2017 (Washington, DC: World Resources Institute), available online at cait.wri.org.

10. Electricity includes energy used by combined heat and power plants and heat plants.

11. Other fuel combustion includes CH_4 and N_2O from biomass combustion, stationary sources, and mobile sources. It also includes CO_2 from commercial and institutional activities, residential, and agriculture, forestry, and fishing. For more information, see "CAIT Country Greenhouse Gas Emissions: Sources and Methods."

12. "9.6.1 Policies Aimed at Reducing Deforestation—AR4 WGIII Chapter 9: Forestry" (Intergovernmental Panel on Climate Change), 9.6.1, accessed December 11, 2017, http://www.ipcc.ch/publications_and_data/ar4/wg3/en/ch9s9-6-1.html.

Chapter 2

1. M. Grubb, C. Carraco, and J. Schellnhuber, "Technological Change for Atmospheric Stabilization: Introductory Overview to the Innovation Modeling Comparison Project," *The Energy Journal* Special Issue, "Endogenous Technological Change and the Economics of Atmospheric Stabilization" (2006): 1–16.

2. "Zero Emission Vehicles Need to Take Over Car Market to Reach 1.5°C Limit:

Analysis" (Climate Action Tracker, 2016), http://climateactiontracker.org/news/260/Zero -emission-vehicles-need-to-take-over-car-market-to-reach-1.5C-limit-analysis.html.

3. Readers interested in more discussion are directed to Jeffrey Rissman, "It Takes a Portfolio: A Broad Spectrum of Policies Can Best Halt Climate Change in Electricity Policy" (Electricity Policy, 2016), http://energyinnovation.org/wp-content/uploads/2018/01/2016 -08-18-Broad-Spectrum-Published-Article.pdf; Chris Busch and Hal Harvey, "Climate Policy for the Real World" (Energy Innovation, 2016), 11–21; and William H. Golove and Joseph H. Eto, "Market Barriers to Energy Efficiency: A Critical Reappraisal of the Rationale for Public Policies to Promote Energy Efficiency" (Lawrence Berkeley National Laboratory, 1996).

4. "Experience Curves for Energy Technology Policy" (International Energy Agency; Organisation for Economic Co-operation and Development, 2000), http://www.wenergy.se /pdf/curve2000.pdf.

5. Zachary Shahan, "13 Charts on Solar Panel Cost & Growth Trends" (CleanTechnica, 2014), https://cleantechnica.com/2014/09/04/solar-panel-cost-trends-10-charts/.

6. Ibid.

7. "Top Runner Program: Developing the World's Best Energy-Efficient Appliance and More" (Japanese Ministry of Economy, Trade, and Industry, 2015), 6–9, http://www.enecho .meti.go.jp/category/saving_and_new/saving/data/toprunner2015e.pdf.

8. United States, "40 CFR 86.1803-01 - Definitions" (n.d.), https://www.law.cornell.edu /cfr/text/40/86.1803-01.

9. Alexis C. MADRIGA, "Why Crossovers Conquered the American Highway," *The Atlantic*, 2014, http://www.theatlantic.com/technology/archive/2014/07/how-the-crossover -conquered-americas-automobile-market/374061/.

10. Arthur van Benthem, Kenneth Gillingham, and James Sweeney, "Learning-by-Doing and the Optimal Solar Policy in California," *The Energy Journal* 29, no. 3 (July 1, 2008), https://doi.org/10.5547/ISSN0195-6574-EJ-Vol29-No3-7.

11. Warren Lasher, "The Competitive Renewable Energy Zone Process" (ERCOT, 2014), http://energy.gov/sites/prod/files/2014/08/f18/c_lasher_qer_santafe_presentation.pdf.

Chapter 3

1. The way we use land can increase or decrease overall emissions as well, but land use is beyond the scope of this book.

2. "Greenhouse Gas Abatement Cost Curves: Sustainability & Resource Productivity," McKinsey & Company, 89, accessed December 11, 2017, https://www.mckinsey.com/busi ness-functions/sustainability-and-resource-productivity/our-insights/greenhouse-gas -abatement-cost-curves.

3. Ibid., 89.

4. The Energy Policy Simulator includes nonenergy policies, such as those affecting land use and industrial processes, as well as energy policies.

5. For more information, see "Energy Policy Simulator Documentation" (Energy Innovation, January 10, 2018), https://us.energypolicy.solutions/docs/.

6. The EPS calculates first-order costs and savings (which entities pay which other entities more or less money relative to the business-as-usual case). It does not include a full macroeconomic simulation of the economy, which would consider questions such as how government spends increased tax revenues (or what government cuts in response to reduced tax revenues), because these decisions are beyond the scope of the energy and nonenergy policies that the computer model seeks to evaluate.

7. The policy cost curve uses a net present value of costs and savings through 2050 and an average annual abatement potential by taking the sum of all future year emission reductions and dividing by the number of years modeled. Results in any specific year might differ substantially.

8. The Energy Innovation team has developed policy cost curves for China, the United States, Mexico, Poland, and Indonesia, and will conclude several more in the next two years. These cover more than half of global carbon emissions, and several can be used as surrogates in other countries.

9. For example, see David Roland-Holst, "Real Incomes, Employment, and California Climate Policy" (Next 10, 2010), 14.

10. Daniel M. Kammen, Kamal Kapadia, and Matthias Fripp, "Putting Renewables to Work: How Many Jobs Can the Clean Energy Industry Generate?" (RAEL Report, University of California, Berkeley, 2004).

11. Tim Buckley and Simon Nicholas, "China's Global Renewable Energy Expansion" (Institute for Energy Economics and Financial Analysis, 2017), http://ieefa.org/wp-content /uploads/2017/01/Chinas-Global-Renewable-Energy-Expansion_January-2017.pdf.

12. Jamie Hosking, Pierpaolo Mudu, and Carlos Dora, "Health Co-Benefits of Climate Change Mitigation," *Health in the Green Economy* (World Health Organization, 2011).

13. U.S. EPA, "Criteria Air Pollutants," Policies and Guidance (U.S. EPA, April 9, 2014), https://www.epa.gov/criteria-air-pollutants.

14. Hosking et al., "Health Co-Benefits of Climate Change Mitigation."

Section I

1. Elmar Kriegler et al., "What Does the 2°C Target Imply for a Global Climate Agreement in 2020? The LIMITS Study on Durban Platform Scenarios" (Data downloaded from the LIMITS database hosted at IIASA), https://tntcat.iiasa.ac.at/LIMITSDB/dsd?Action= htmlpage&page=about.

2. Ibid.

Chapter 4

1. World Resources Institute, "CAIT Climate Data Explorer" (2017), http://cait.wri.org.

2. Toby Couture, Karlynn Cory, Claire Kreycik, and Emily Williams, "A Policymaker's Guide to Feed-In Tariff Policy Design" (National Renewable Energy Laboratory, 2010), vi, http://www.nrel.gov/docs/fy10osti/44849.pdf.

3. Load-serving entities (LSEs) are the providers of electricity for retail customers. They include utilities and retail electricity providers.

4. For example, see "PSEG Long Island: Commercial Solar PV Feed-In Tariff" (DSIRE, 2017), http://programs.dsireusa.org/system/program/detail/21865; "Clean Energy Standard" (DSIRE, 2017), http://programs.dsireusa.org/system/program/detail/5883.

5. Matthias Lang and Annette Lang, "German Feed-In Tariffs 2014 (from 08)," *German Energy Blog* (blog), accessed December 12, 2017, http://www.germanenergyblog.de/?page _id=16379.

6. Fraunhofer Institute, "Energy Charts," accessed December 12, 2017, https://www.en ergy-charts.de/power_inst.htm.

7. "Global Offshore Wind Capacity Reaches 14.4GW in 2016," *Offshore Wind*, accessed December 12, 2017, https://www.offshorewind.biz/2017/02/10/global-offshore-wind-ca pacity-reaches-14-4gw-in-2016/.

8. Timothy Jones, "German Offshore Wind Park to Be Built without Subsidies," *Deutsche*

Welle (2017), http://www.dw.com/en/german-offshore-wind-park-to-be-built-without-subsidies/a-38430493.

9. "Renewable Energy Sources Act (Erneuerbare-Energien-Gesetz EEG)" (International Energy Agency, 2004), https://www.iea.org/policiesandmeasures/pams/germany/name-22369-en.php.

10. Ibid.

11. Fraunhofer Institute, "Energy Charts."

12. Kerstine Appunn, "EEG Reform 2016: Switching to Auctions for Renewables," *Clean Energy Wire* (2016), https://www.cleanenergywire.org/factsheets/eeg-reform-2016-switching-auctions-renewables.

13. See Trieu Mai, David Mulcahy, M. Maureen Hand, and Samuel F. Baldwin, "Envisioning a Renewable Electricity Future for the United States," *Energy* 65 (2014): 374–86, https://doi.org/10.1016/j.Energy.2013.11.029; "Investigating a Higher Renewables Portfolio Standard in California" (Energy and Environmental Economics, Inc., 2014), https://www.ethree.com/wp-content/uploads/2017/01/E3_Final_renewable portfolio standard_Report_2014_01_06_with_appendices.pdf; "The Low Carbon Grid Study" (California 2030 Low Carbon Grid Study, 2017), http://lowcarbongrid2030.org/; and Alexander E. MacDonald et al., "Future Cost-Competitive Electricity Systems and Their Impact on U.S. CO_2 Emissions," *Nature Climate Change* 6 (January 25, 2016): 526.

14. "International Energy Outlook 2016" (U.S. Energy Information Administration, 2016), https://www.eia.gov/outlooks/ieo/ieo_tables.php.

15. Jesse Jenkins, "What's Killing Nuclear Power in U.S. Electricity Markets?" (MIT Center for Energy and Environmental Policy Research, 2017), http://ceepr.mit.edu/files/papers/2018-001-Brief.pdf.

16. Of course, the compliance and renewable energy credit tracking regimes should be synced to prevent double counting and ensure the renewable portfolio standard policy is having its intended effect of driving new investment in renewable generation.

17. For more detailed information on the pitfalls of using tax credits, see Chapter 6.

18. For more information on U.S. renewable energy credit tracking systems, see "Renewable Energy Certificate Claims and Ownership" (Center for Resource Solutions, 2015), https://resource-solutions.org/learn/rec-claims-and-ownership/.

19. Galen Barbose, "U.S. Renewables Portfolio Standards 2016 Annual Status Report," Slide 5, Lawrence Berkeley National Laboratory, 2016. https://emp.lbl.gov/sites/all/files/lbnl-1005057.pdf.

20. "Electricity Data Browser" (U.S. Energy Information Administration, December 12, 2017), https://www.eia.gov/electricity/data/browser/.

21. Barbose, "U.S. Renewables Portfolio Standards 2016 Annual Status Report," Slide 34.

22. Technically these resources can be built out of state if they are connected to the state's grid.

23. "2016 Detailed Electricity Statistics" (China Energy Portal, 2017), https://chinaenergyportal.org/2016-detailed-electricity-statistics/.

24. 49.7 TWh was curtailed, compared with a total of 241 TWh generated. Brian Publicover, "China Adds 19.3GW of Wind in 2016 but Curtailment Soars," *Recharge News* (2017), http://www.rechargenews.com/wind/1210416/china-adds-193gw-of-wind-in-2016-but-curtailment-soars.

25. John Parnell, "China Trials Wind and Solar Certificate Scheme in Move Away from Feed-In Tariffs," *PV Tech* (2017), https://www.pv-tech.org/news/china-trials-wind-and-solar-certificate-scheme-in-move-away-from-feed-in-ta.

26. Nagalakshmi Puttaswamy and Mohd. Sahil Ali, "How Did China Become the Largest Solar PV Manufacturing Country?" (Center for Study of Science, Technology, & Policy, 2015), http://www.cstep.in/uploads/default/files/publications/stuff/CSTEP_Solar_PV_Working_Series_2015.pdf.

27. Christian Redl, "The Recent Revision of Renewable Energy Act in Germany: Overview and Results of the PV Tendering Scheme" (Agora Energiewende, 2016), http://rekk.hu/downloads/events/2017_SEERMAP_Redl_EEG%20and%20tendering_Sofia.pdf.

Chapter 5

1. Trieu Mai et al., "Envisioning a Renewable Electricity Future for the United States," *Energy* 65, Supplement C (February 1, 2014): 374–86, https://doi.org/10.1016/j.energy.2013.11.029; Alexander E. MacDonald et al., "Future Cost-Competitive Electricity Systems and Their Impact on US CO_2 Emissions," *Nature Climate Change* 6 (January 25, 2016): 526.

2. Sonia Aggarwal, "Clean Energy, Batteries Not Included (Op-Ed)," *Live Science* (blog), 2014, https://www.livescience.com/46973-clean-energy-storage-without-using-batteries.html.

3. Stefan M. Knupfer et al., "Electrifying Insights: How Automakers Can Drive Electrified Vehicle Sales and Profitability" (McKinsey & Company, 2017), https://www.mckinsey.com/industries/automotive-and-assembly/our-insights/electrifying-insights-how-automakers-can-drive-electrified-vehicle-sales-and-profitability.

4. Jeff St. John, "UK's National Grid Goes Big into Energy Storage with 201MW of Fast-Acting Batteries" *Greentech Media* (2016), https://www.greentechmedia.com/articles/read/uks-national-grid-goes-big-into-energy-storage-with-201mw-of-fast-acting-ba.

5. Sonia Aggarwal and Jeffrey Gu, "Two Kinds of Demand Response" (Energy Innovation, 2012), http://energyinnovation.org/wp-content/uploads/2014/11/Two-Kinds-of-Demand-Response.pdf.

6. Sonia Aggarwal and Michael O'Boyle, "Getting the Most from Grid Modernization" (Energy Innovation, 2017), http://energyinnovation.org/wp-content/uploads/2017/02/GridModernizationMetricsOutcomes_Feb2017.pdf.

7. For a deep dive on the capital biases created by cost of service regulation, see Steve Kihm et al., "You Get What You Pay For: Moving toward Value in Utility Compensation, Part 1: Revenue and Profit" (America's Power Plan, 2015), http://americaspowerplan.com/wp-content/uploads/2016/07/CostValue-Part1-Revenue.pdf; and Dan Aas and Michael O'Boyle, "You Get What You Pay For: Moving toward Value in Utility Compensation, Part 2: Regulatory Alternatives" (America's Power Plan, 2016), http://americaspowerplan.com/wp-content/uploads/2016/08/2016_Aas-OBoyle_Reg-Alternatives.pdf.

8. A host of resources on designing performance-based regulation can be found at "Going Deep on Performance-Based Regulation" (Energy Innovation: Policy and Technology, 2016), http://energyinnovation.org/publication/going-deep-performance-based-regulation/.

9. Aas and O'Boyle, "You Get What You Pay For: Moving toward Value in Utility Compensation."

10. Ron Lehr, "New Utility Business Models: Utility and Regulatory Models for the Modern Era" (America's Power Plan, 2013), http://americaspowerplan.com/wp-content/uploads/2013/10/APP-UTILITIES.pdf.

11. Seth Nowak et al., "Beyond Carrots for Utilities: A National Review of Performance Incentives for Energy Efficiency" (American Council for an Energy-Efficient Economy, 2015), https://aceee.org/beyond-carrots-utilities-national-review.

12. For more information on how regional transmission organizations function, see "Regional Transmission Organizations: The Basics" (Western Clean Energy Advocates, 2016), http://westerngrid.net/wcea/rso/.

13. Mike Hogan, "Power Markets: Aligning Power Markets to Deliver Value" (America's Power Plan, 2013), http://americaspowerplan.com/wp-content/uploads/2014/01/APP-Markets-Paper.pdf.

14. For detailed examples of how wholesale market operators are adapting to value flexibility, see Robbie Orvis and Sonia Aggarwal, "A Roadmap for Finding Flexibility in Wholesale Markets" (America's Power Plan; Energy Innovation, 2017).

15. Brendan Pierpont et al., "Flexibility: The Path to Low-Carbon, Low-Cost Electricity Grids" (Climate Policy Initiative, 2017), https://climatepolicyinitiative.org/publication/flexibility-path-low-carbon-low-cost-electricity-grids/.

16. Sonia Aggarwal, "Future Wholesale Markets and Implications for Retail Markets" (America's Power Plan, 2017), http://energyinnovation.org/wp-content/uploads/2017/09/20170914_WholesaleMarkets-SoniaAggarwal.pdf.

17. Jairo Chung et al., "Rate-Basing Wind Generation Adds Momentum to Renewables" (Moody's, 2017), https://www.researchpool.com/provider/moodys-investors-service/us-power-and-utilities-rate-basing-wind-generation-adds-momentum-to-r/.

18. Eric Gimon, "New Financial Tools Proposed in Colorado Could Solve Coal Retirement Conundrum," *Forbes* (2017), https://www.forbes.com/sites/energyinnovation/2017/04/19/new-financial-tools-proposed-in-colorado-could-solve-coal-retirement-conundrum/#76c34c2311c5.

19. Robbie Orvis, "The State of US Wholesale Power Markets: Is Reliability at Risk from Low Prices?," *Utility Dive* (2017), https://www.utilitydive.com/news/the-state-of-us-wholesale-power-markets-is-reliability-at-risk-from-low-pr/443273/.

20. Robbie Orvis and Eric Gimon, "The State of Wholesale Power Markets: Principles for Managing an Evolving Power Mix" *Utility Dive* (2017), https://www.utilitydive.com/news/the-state-of-wholesale-power-markets-principles-for-managing-an-evolving-p/447839/.

21. Travis Houm, "Forget Tax Credits. These Two Federal Policy Changes Are a More Immediate Threat to Solar" *Greentech Media* (2017), https://www.greentechmedia.com/articles/read/these-two-federal-policy-changes-could-be-way-more-damaging-to-solar.

22. Let's say you have a solar project that has predictable, contracted cash flows of $1 million per year for the next 25 years. If the rate of return investors demand is 7 percent, then the project is worth $11.65 million. But if the rate of return rises just one percentage point to 8 percent, the value of the project falls 8.4 percent to $10.67 million. A rise to 9 percent reduces the value by 15.7 percent to $9.82 million.

23. O. Zinaman, A.M. Miller, and D. Aent, "Power Systems of the Future" (National Renewable Energy Laboratory [NREL], 2015).

24. Pierpont et al., "Flexibility."

25. It may be that utilities, which often own and operate these plants, are not reliable sources of information on the viability of retirement. Therefore, objective third-party analysis may be useful, and international organizations such as the 21st Century Power Partnership can help bolster analytical capacity. See https://www.21stcenturypower.org/.

26. "Energy Efficiency: A Compelling Global Resource" (McKinsey & Company, 2010), http://www.mckinsey.com/~/media/mckinsey/dotcom/client_service/sustainability/pdfs/a_compelling_global_resource.ashx.

27. Justin Guay et al., "Expanding Energy Access beyond the Grid: Five Principles for Designing Off-Grid and Mini-Grid Policy" (Sierra Club, 2014), https://www.sierraclub.org

/sites/www.sierraclub.org/files/uploads-wysiwig/Expanding%20Energy%20Access%20
Beyond%20The%20Grid%20Report%202014.pdf.

28. Ibid.

29. Ron Binz, et al., "Practicing Risk-Aware Public Utility Regulation" (Ceres, April 2012), http://www.raponline.org/wp-content/uploads/2016/05/ceres-binzsedano-riskawareregulation-2012-apr-19.pdf.

30. Ryan Wiser and Mark Bolinger, "2016 Wind Technologies Market Report" (U.S. Department of Energy, 2016), https://energy.gov/sites/prod/files/2017/10/f37/2016_Wind_Technologies_Market_Report_101317.pdf. Data taken from Office of Energy Efficiency & Renewable Energy, "2016 Wind Technologies Market Report," https://energy.gov/eere/wind/downloads/2016-wind-technologies-market-report.

31. Samuel A. Newell, et al., "Pricing Carbon into NYISO's Wholesale Energy Market to Support New York's Decarbonization Goals" (The Brattle Group, 2017), http://www.nyiso.com/public/webdocs/markets_operations/documents/Studies_and_Reports/Studies/Market_Studies/Pricing_Carbon_into_NYISOs_Wholesale_Energy_Market.pdf.

32. Clyde Loutan et al., "Demonstration of Essential Reliability Services by a 300-MW Solar Photovoltaic Power Plant" (National Renewable Energy Laboratory, 2017), https://www.nrel.gov/docs/fy17osti/67799.pdf.

33. Melissa Whited, Tim Woolf, and Alice Napolean, "Utility Performance Incentive Mechanisms: A Handbook for Regulators" (Synapse Energy Economics, 2015), http://www.synapse-energy.com/sites/default/files/Utility%20Performance%20Incentive%20Mechanisms%2014-098_0.pdf.

34. Michael O'Boyle, "Designing a Performance Incentive Mechanism for Peak Load Reduction: A Straw Proposal" (Energy Innovation, 2016), http://americaspowerplan.com/wp-content/uploads/2014/10/Peak-Reduction-PIM-whitepaper.pdf.

35. Whited, et al., "Utility Performance Incentive Mechanisms."

36. Michael Milligan, "Capacity Value of Wind Plants and Overview of U.S. Experience" (Stockholm, Sweden, 2011), https://www.nrel.gov/docs/fy11osti/52856.pdf.

37. Andrew Mills and Ryan Wiser, "Implications of Wide-Area Geographic Diversity for Short-Term Variability of Solar Power" (National Renewable Energy Laboratory, 2010), https://escholarship.org/uc/item/9mz3w055.

38. "Quarterly Benefits and Greenhouse Gas Reports" (Western Energy Imbalance Market, 2017), accessed December 14, 2017, https://www.westerneim.com/Pages/About/QuarterlyBenefits.aspx.

39. Andrew Liu et al., "Co-Optimization of Transmission and Other Supply Resources" (Eastern Interconnection States' Planning Council; National Association of Regulatory Utility Commissioners, 2013), https://pubs.naruc.org/pub.cfm?id=536D834A-2354-D714-51D6-AE55F431E2AA.

40. MacDonald, et al., "Future Cost-Competitive Electricity Systems and Their Impact on U.S. CO_2 Emissions."

41. For a more detailed discussion of performance metrics for publicly owned utilities, see Michael O'Boyle and Sonia Aggarwal, "Improving Performance in Publicly-Owned Utilities" (America's Power Plan, 2015), http://americaspowerplan.com/wp-content/uploads/2015/12/ImprovingPerformancePubliclyOwnedUtilities.pdf.

42. For a fuller picture of decoupling and its design options, see Jim Lazar, Frederick Weston, and Wayne Shirley, "Revenue Regulation and Decoupling: A Guide to Theory and Application, Second Printing" (Regulatory Assistance Project, November 2016), http://www

.raponline.org/wp-content/uploads/2016/11/rap-revenue-regulation-decoupling-guide
-second-printing-2016-november.pdf.

43. Gimon, "New Financial Tools Proposed in Colorado Could Solve Coal Retirement Conundrum."

44. Robbie Searcy, "ERCOT Breaks Peak Record Again, Tops 71,000 MW for First Time," *ERCOT News Releases* (blog), August 11, 2016, http://www.ercot.com/news/releases/show /103663.

45. Mark Watson, "Wind Hits New Monthly Record for Share of ERCOT Energy" (*Platts*, 2017), https://www.platts.com/latest-news/electric-power/houston/wind-hits-new-monthly -record-for-share-of-ercot-21410272.

46. Herman Trabish, "Texas CREZ Lines Delivering Grid Benefits at $7B Price Tag," *Utility Dive* (2014), https://www.utilitydive.com/news/texas-crez-lines-delivering-grid-benefits -at-7b-price-tag/278834/.

47. "ERCOT Generator Interconnection Status Report" (ERCOT, 2017), http://www .ercot.com/content/wcm/lists/114799/GIS_REPORT__October_2017.xlsx.

48. Ibid.

49. "A Wind Vision for New Growth in Texas" (Wind Energy Foundation; American Wind Energy Association, 2015), http://awea.files.cms-plus.com/TEXAS%20REPORT_11 _16_15.pdf.

50. Eric Gimon, "Texas Regulators Saved Customers Billions by Avoiding a Traditional Capacity Market" *Greentech Media* (2016), https://www.greentechmedia.com/articles /read/texas-regulators-save-customers-billions.

51. Robbie Searcy, "ERCOT Reports Indicate Enough Generation Resources for Summer, Coming Years" (ERCOT, 2017), http://www.ercot.com/news/releases/show /123359.

52. Paul Noothout, et al., "The Impact of Risks in Renewable Energy Investments and the Role of Smart Policies" (ECOFYS, 2016), 40, http://diacore.eu/images/files2/WP3 -Final%20Report/diacore-2016-impact-of-risk-in-res-investments.pdf.

53. Ibid., 112.

54. David Nelson, et al., "Policy and Investment in German Renewable Energy" (Climate Policy Initiative, 2016), https://climatepolicyinitiative.org/wp-content/uploads/2016/04 /Policy-and-investment-in-German-renewable-energy.pdf.

55. "2014 Amendment of the Renewable Energy Sources Act (EEG 2014)" (International Energy Agency, 2014), https://www.iea.org/policiesandmeasures/pams/germany/name -145053-en.php.

56. "Renewable Energy Sources Act (Erneuerbare-Energien-Gesetz EEG)" (International Energy Agency, 2004), https://www.iea.org/policiesandmeasures/pams/germany/name -22369-en.php.

Section II

1. World Resources Institute, "CAIT Climate Data Explorer" (2017), http://cait.wri.org.

2. Elmar Kriegler et al., "What Does the 2°C Target Imply for a Global Climate Agreement in 2020? The LIMITS Study on Durban Platform Scenarios" (Data downloaded from the LIMITS database hosted at IIASA), https://tntcat.iiasa.ac.at/LIMITSDB/dsd?Action= htmlpage&page=about.

3. Ibid.

4. Ibid. Demand measured as passenger-kilometers and freight-ton-kilometers.

Chapter 6

1. World Resources Institute, "CAIT Climate Data Explorer" (2017), http://cait.wri.org.

2. "Pathways to a Low-Carbon Economy: Version 2 of the Global Greenhouse Gas Abatement Cost Curve" (McKinsey & Company, 2009), https://www.mckinsey.com/~/media/McKinsey/Business%20Functions/Sustainability%20and%20Resource%20Productivity/Our%20Insights/Pathways%20to%20a%20low%20carbon%20economy/Pathways%20to%20a%20low%20carbon%20economy.ashx.

3. "Climate Change 2014: Mitigation of Climate Change" (Intergovernmental Panel on Climate Change, 2014), http://www.ipcc.ch/pdf/assessment-report/ar5/wg3/ipcc_wg3_ar5_chapter8.pdf.

4. "The Chinese Automotive Fuel Economy Policy" (United Nations Environment Programme, 2015), https://www.globalfueleconomy.org/transport/gfei/autotool/case_studies/apacific/china/CHINA%20CASE%20STUDY.pdf.

5. David L. Greene, "How Consumers Value Fuel Economy: A Literature Review" (U.S. EPA, n.d.), https://nepis.epa.gov/Exe/ZyPDF.cgi/P1006V0O.PDF?Dockey=P1006V0O.PDF.

6. "New ICAO Aircraft CO_2 Standard One Step Closer to Final Adoption" (ICAO, 2016), https://www.icao.int/Newsroom/Pages/New-ICAO-Aircraft-CO2-Standard-One-Step-Closer-To-Final-Adoption.aspx.

7. "Adoption of an Energy Efficiency Design Index for International Shipping" (U.S. EPA, 2011), https://nepis.epa.gov/Exe/ZyPDF.cgi/P100BK43.PDF?Dockey=P100BK43.PDF.

8. U.S. Energy Information Administration, "Monthly Energy Review," accessed February 5, 2018, https://www.eia.gov/totalenergy/data/monthly/.

9. Bureau of Transportation Statistics, "U.S. Vehicle-Miles (Millions)," accessed February 5, 2018, https://www.bts.gov/content/us-vehicle-miles-millions.

10. "Fuel Economy: Going Farther on a Gallon of Gas" (Union of Concerned Scientists, 2003), http://lobby.la.psu.edu/_107th/126_CAFE_Standards_2/Organizational_Statements/UCS/UCS_Fuel_Economy_Going_Farther.htm.

11. Osamu Kimura, "Japanese Top Runner Approach for Energy Efficiency Standards," *SERC Discussion Paper*, no. SERC09035 (2010), https://www.researchgate.net/publication/228900679_Japanese_Top_Runner_Approach_for_energy_efficiency_standards.

12. Nicholas Lutsey, "New Automobile Regulations: Double the Fuel Economy, Half the CO_2 Emissions, and Even Automakers Like It," *ACCESS Magazine* (2012), https://www.accessmagazine.org/fall-2012/new-automobile-regulations/.

13. Jeffrey Rissman and Maxine Savitz, "Unleashing Private-Sector Energy R&D: Insights from Interviews with 17 R&D Leaders" (Energy Innovation Council, 2013), http://energyinnovation.org/wp-content/uploads/2014/06/unleashing-private-rd-jan2013.pdf.

14. Peter Mock, "Vehicle Emissions Testing in the EU: Why We Are Still Struggling with the Dead Hand of the Past—and What the Future Is Likely to Bring" (International Council on Clean Transportation, 2015), http://www.theicct.org/blogs/staff/vehicle-co2-testing-eu-still-struggling.

15. Ibid.

16. "Fuel Efficiency Standards for Vehicles—Top Runner Program" (International Energy Agency, 1979), http://www.iea.org/policiesandmeasures/pams/japan/name-24367-en.php.

17. Kimura, "Japanese Top Runner Approach for Energy Efficiency Standards."

18. "Top Runner Program: Developing the World's Best Energy-Efficient Appliance and More" (Japanese Ministry of Economy, Trade, and Industry, 2015), 13, http://www.enecho.meti.go.jp/category/saving_and_new/saving/data/toprunner2015e.pdf.

19. "Japan: Light Duty: Fuel Economy" (Transport Policy), accessed December 14, 2017, http://www.transportpolicy.net/standard/japan-light-duty-fuel-economy/.

20. Ibid., 6.

21. Ibid., 34.

22. Ibid., 9.

23. "China: Light-Duty: Fuel Consumption" (Transport Policy), accessed December 14, 2017, http://www.transportpolicy.net/standard/china-light-duty-fuel-consumption/.

24. "The Chinese Automotive Fuel Economy Policy," 3.

25. Ibid.

26. Ibid., 13–16.

27. Ibid., 19.

Chapter 7

1. Francisco Posada Sanchez and Laura Segafredo, "Policies That Work: How Vehicle Standards and Fuel Fees Can Cut CO_2 Emissions and Boost the Economy" (ClimateWorks Foundation, 2012), 5, https://cleanenergysolutions.org/sites/default/files/documents/CW-ICCT-PTW-CESC-December-13-2012_Final.pdf.

2. Anastasia Kharina, Daniel Rutherford, and Mazyar Zeinali, "Cost Assessment of Near- and Mid-Term Technologies to Improve New Aircraft Fuel Efficiency" (International Council on Clean Transportation, 2016), http://www.theicct.org/publications/cost-assessment-near-and-mid-term-technologies-improve-new-aircraft-fuel-efficiency.

3. "State Motor Fuel Taxes: Notes Summary" (American Petroleum Institute, 2016), http://www.api.org/~/media/Files/Statistics/State-Motor-Fuel-Excise-Tax-Update-July-2016.pdf.

4. "2017 and Later Model Year Light-Duty Vehicle Greenhouse Gas Emissions and Corporate Average Fuel Economy Standards," 40 CFR Parts 85, 86, and 600 § (2012), 62716, http://www.gpo.gov/fdsys/pkg/FR-2012-10-15/pdf/2012-21972.pdf.

5. "Greenhouse Gas Emissions Standards and Fuel Efficiency Standards for Medium- and Heavy-Duty Engines and Vehicles," 49 CFR Parts 523, 534, and 535 § (2011), 57329, http://www.gpo.gov/fdsys/pkg/FR-2011-09-15/pdf/2011-20740.pdf.

6. "In China, the License Plates Can Cost More Than the Car" *Bloomberg* (2013), https://www.bloomberg.com/news/articles/2013-04-25/in-china-the-license-plates-can-cost-more-than-the-car.

7. Leo Mirani, "Norway's Electric-Car Incentives Were So Good They Had to Be Stopped," *Quartz* (blog), accessed December 14, 2017, https://qz.com/400277/norway-electric-car-incentives-were-so-good-they-had-to-be-stopped/.

8. Peter Levring, "Soon, Cars in Denmark Will Only Be Taxed at 100%" *Bloomberg* (2017), https://www.bloomberg.com/news/articles/2017-08-29/soon-cars-in-denmark-will-only-be-taxed-at-100.

9. Peter Levring, "Denmark Is Killing Tesla (and Other Electric Cars)" *Bloomberg* (2017), https://www.bloomberg.com/news/articles/2017-06-02/denmark-is-killing-tesla-and-other-electric-cars.

10. Meghan R. Busse, Christopher R. Knittel, and Florian Zettelmeyer, "Are Consumers Myopic? Evidence from New and Used Car Purchases," *American Economic Review* 103, no. 1 (2013): 220–56, https://doi.org/10.1257/aer.103.1.220.

11. Additional tools can help policymakers design the right feebate structure for their countries. For example, see "Feebate Simulation Tool" (International Council on Clean Transportation), accessed December 14, 2017, http://www.theicct.org/feebate-simulation-tool.

12. David Greene, "What Is Greener Than a VMT Tax? The Case for an Indexed Energy

User Fee to Finance US Surface Transportation," *Transportation Research Part D: Transport and Environment* 16 (August 1, 2011): 451–58, https://doi.org/10.1016/j.trd.2011.05.003.

13. Vehicle performance standards that are set by vehicle weight rather than footprint risk incentivizing manufacturers to make heavier vehicles so they can be categorized in a heavier vehicle class with lower efficiency requirements.

14. Kevin McCormally, "A Brief History of the Federal Gasoline Tax" (Kiplinger, 2014), http://www.kiplinger.com/article/spending/T063-C000-S001-a-brief-history-of-the-federal -gasoline-tax.html.

15. Ibid.

16. Brad Plumer, "A Short History of America's Gas Tax Woes," *Washington Post* (2011), https://www.washingtonpost.com/blogs/wonkblog/post/a-short-history-of-americas-gas -tax-woes/2011/08/24/gIQAjyfXdJ_blog.html.

17. Bennett Cohen and Cory Lowe, "Feebates: A Key to Breaking U.S. Oil Addiction" (Rocky Mountain Institute, 2010), https://www.rmi.org/news/feebates-key-breaking -u-s-oil-addiction/.

18. "Bonus-Malus: Definitions and Scales for 2016" (French Ministry of Environment, Energy, and the Sea, 2016), http://www.developpement-durable.gouv.fr/Bonus-Malus -definitions-et-baremes.

19. John German and Dan Meszler, "Best Practices for Feebate Program Design and Implementation" (International Council on Clean Transportation, 2010), http://www.theicct .org/sites/default/files/publications/ICCT_feebates_may2010.pdf.

20. "Bonus-Malus Écologique" (Wikipédia, 2017), https://fr.wikipedia.org/w/index .php?title=Bonus-malus_%C3%A9cologique&oldid=143338453.

Chapter 8

1. "All-Electric Vehicles" (U.S. Department of Energy), accessed December 15, 2017, http://www.fueleconomy.gov/feg/evtech.shtml.

2. Some emissions are associated with building power plants and transmission lines, mining uranium for nuclear plants, and methane emissions from reservoirs created by hydroelectric dams.

3. "How Do Gasoline and Electric Vehicles Compare?" (Idaho National Laboratory, n.d.), https://avt.inl.gov/sites/default/files/pdf/fsev/compare.pdf.

4. This figure is based on an EV efficiency of 3 miles/kWh and an electricity cost of 7 cents/kWh.

5. Erin Stepp, "Your Driving Costs" (AAA NewsRoom, 2016), http://newsroom.aaa.com /auto/your-driving-costs/.

6. "How Do Gasoline and Electric Vehicles Compare?"

7. *Edison Electric Bus from 1915*, n.d., Photograph from Wikimedia Commons, n.d., Library of Commons, https://commons.wikimedia.org/wiki/File:Edison_electric_bus_from _1915.jpg.

8. James Ayre, "China 100% Electric Bus Sales Grew To ~115,700 in 2016" *CleanTechnica* (2017), https://cleantechnica.com/2017/02/03/china-100-electric-bus-sales-grew-115700 -2016/.

9. "Shenzhen Completes Switch to Fully Electric Bus Fleet. Electric Taxis Are Next" *CleanTechnica* (2018), https://cleantechnica.com/2018/01/01/shenzhen-completes-switch -fully-electric-bus-fleet-electric-taxis-next/.

10. Fred Lambert, "Electric Buses Are Now Cheaper Than Diesel/CNG and Could

Dominate the Market within 10 Years, Says Proterra CEO" *Electrek* (2017), https://electrek
.co/2017/02/13/electric-buses-proterra-ceo/.

11. Fred Lambert, "Tesla Semi: Analysts See Tesla Leasing Batteries for $0.25/Miles in
300,000 Electric Trucks for $7.5 Billion in Revenue" *Electrek* (2017), https://electrek.co/2017
/04/20/tesla-semi-leasing-batteries-electric-truck/.

12. Martin R. Cohen, "The ABCs of EVs: A Guide for Policy Makers and Consumer Advocates" (Citizens Utility Board, 2017), https://citizensutilityboard.org/wp-content/uploads
/2017/04/2017_The-ABCs-of-EVs-Report.pdf.

13. Björn Nykvist and Måns Nilsson, "Rapidly Falling Costs of Battery Packs for
Electric Vehicles," *Nature Climate Change* 5 (2015): 329, https://www.nature.com/articles
/nclimate2564; and Steve Hanley, "Electric Vehicle Battery Prices Are Falling Faster Than
Expected" *CleanTechnica* (2017), https://cleantechnica.com/2017/02/13/electric-vehicle
-battery-prices-falling-faster-expected/.

14. "Where to Use" (Bay Area FasTrak), accessed December 15, 2017, https://www
.bayareafastrak.org/en/howitworks/whereToUse.shtml#bay.

15. "Estimated Electric Vehicle Charge Times" (Clipper Creek, 2017), https://www.clip
percreek.com/wp-content/uploads/2017/06/TIME-TO-CHARGE-20170612-final-low-res.pdf.

16. Brad Berman, "Quick Charging of Electric Cars" *PluginCars* (2014), http://
www.plugincars.com/electric-car-quick-charging-guide.html.

17. Mark Uh, "From Own to Rent: Who Lost the American Dream?," (Trulia, 2016),
https://www.trulia.com/blog/trends/own-to-rent/.

18. Cohen, "The ABCs of EVs," 15.

19. Ibid., 15.

20. Ibid.

21. Ibid.

22. John Voelcker, "Many Car Dealers Don't Want to Sell Electric Cars: Here's Why,"
Green Car Reports (2014), https://www.greencarreports.com/news/1090281_many-car
-dealers-dont-want-to-sell-electric-cars-heres-why; and Matt Richtel, "A Car Dealers Won't
Sell: It's Electric," *The New York Times*, 2015, sec. Science, https://www.nytimes.com/2015
/12/01/science/electric-car-auto-dealers.html.

23. Cohen, "The ABCs of EVs," 15.

24. Richtel, "A Car Dealers Won't Sell."

25. Cohen, "The ABCs of EVs," 15.

26. Voelcker, "Many Car Dealers Don't Want to Sell Electric Cars."

27. "What Is ZEV?" (Union of Concerned Scientists, 2016), https://www.ucsusa.org
/clean-vehicles/california-and-western-states/what-is-zev.

28. Colin Murphy, NextGen Climate America, unpublished note, 2017.

29. "China Zero Emission Vehicle Requirement Mandate Boosts Battery Electric Pow-
ertrain Demand" (PR Newswire, 2017), https://www.prnewswire.com/news-releases/china
-zero-emission-vehicle-requirement-mandate-boosts-battery-electric-powertrain-demand
-300543992.html.

30. "EU Contemplates Introduction of Minimum Quotas for the Sales of Electric Vehicles" (Bellona.org, 2017), http://bellona.org/news/transport/electric-vehicles/2017-06
-eu-contemplates-introduction-of-minimum-quotas-for-the-sales-of-electric-vehicles.

31. Anmar Frangoul, "Around 2.2 Million Deaths in India and China from Air Pollution:
Study" *CNBC* (2017), https://www.cnbc.com/2017/02/14/around-22-million-deaths-in
-india-and-china-from-air-pollution-study.html.

32. Michael Caputa, "Georgia EV Sales Sputter without Tax Credit" *Marketplace* (2016), http://www.marketplace.org/2016/01/08/world/georgia-ev-sales-sputter-without-tax-break.

33. "Electric Vehicle Charging Station Permit" (City of Fremont, CA), accessed December 15, 2017, https://fremont.gov/2746/Electric-Vehicle-Charging-Station.

34. "Electric Vehicle Charging System Permit Requirements—SF/Duplexes" (City of San Jose, CA, 2016), https://www.sanjoseca.gov/DocumentCenter/View/1825.

35. Loren McDonald, "Predicting When US Federal EV Tax Credit Will Expire for Tesla Buyers" *CleanTechnica* (2017), https://cleantechnica.com/2017/01/20/predicting-us-federal-ev-tax-credit-will-expire-tesla-buyers/; and John Voelcker, "When Do Electric-Car Tax Credits Expire? (Further Update)" *Green Car Reports* (2016), https://www.greencarreports.com/news/1085549_when-do-electric-car-tax-credits-expire.

36. Severin Borenstein and Lucas W. Davis, "The Distributional Effects of U.S. Clean Energy Tax Credits" (NBER Tax Policy and the Economy, 2015), https://doi.org/10.3386/w21437.

37. Andria Simmons, "Georgia Slams Brakes on Electric Cars," *Atlanta Journal-Constitution* (2015), http://www.govtech.com/state/Georgia-Slams-Brakes-on-Electric-Cars-.html.

38. Ibid.

39. Cohen, "The ABCs of EVs," 11.

40. Aaron Gould Sheinin, "Electric Vehicle Sales Fizzle after Georgia Pulls Plug on Tax Break," *Atlanta Journal-Constitution* (2015), http://www.myajc.com/news/state--regional-govt--politics/electric-vehicle-sales-fizzle-after-georgia-pulls-plug-tax-break/HC0We2aNiekLEn6VoNsa6M/.

41. Matt Smith, "The Georgia Legislature Just Pulled the Plug on Electric Cars," *VICE News* (2015), https://news.vice.com/article/the-georgia-legislature-just-pulled-the-plug-on-electric-cars.

42. Ibid.

43. Simmons, "Georgia Slams Brakes on Electric Cars."

44. Caputa, "Georgia EV Sales Sputter without Tax Credit."

45. Sheinin, "Electric Vehicle Sales Fizzle after Georgia Pulls Plug on Tax Break."

46. Keith Bradsher, "China View to Be World's Leader in Electric Cars," *The New York Times* (2009), http://www.nytimes.com/2009/04/02/business/global/02electric.html.

47. Ibid.

48. Ibid.

49. Jake Spring, "China's Anti-Teslas: Cheap Models Drive Electric Car Boom" *Reuters* (2017), https://www.reuters.com/article/us-usa-autoshow-china-electric/chinas-anti-teslas-cheap-models-drive-electric-car-boom-idUSKBN14V1H3.

50. "Global EV Outlook 2016" (International Energy Agency, 2016), 17, https://www.iea.org/publications/freepublications/publication/Global_EV_Outlook_2016.pdf.

51. Ibid.

52. "Global EV Outlook 2016," 20.

53. Michael J. Dunne, "China Deploys Aggressive Mandates to Take Lead in Electric Vehicles," *Forbes* (2017), https://www.forbes.com/sites/michaeldunne/2017/02/28/china-deploys-aggressive-mandates-to-stay-no-1-in-electric-vehicles/.

54. "Tougher Rules for Electric Vehicle Subsidies," *China Daily* (2017), sec. Business, http://www.chinadaily.com.cn/business/motoring/2017-01/03/content_27842725.htm.

55. Hongyang Cui, "Subsidy Fraud Leads to Reforms for China's EV Market"

(International Council on Clean Transportation, 2017), http://www.theicct.org/blogs/staff/subsidy-fraud-reforms-china-ev-market.

56. "Tougher Rules for Electric Vehicle Subsidies."

57. Spring, "China's Anti-Teslas."

58. "Tougher Rules for Electric Vehicle Subsidies."

59. Mark Kane, "China Considers ZEV Mandate Similar to California" *Inside EVs* (2014), https://insideevs.com/china-considers-zev-mandate/.

60. Spring, "China's Anti-Teslas."

Chapter 9

1. "Open-Air Computers," *The Economist* (2012), https://www.economist.com/news/special-report/21564998-cities-are-turning-vast-data-factories-open-air-computers.

2. C.C. Huang et al., "12 Green Guidelines: CDBC's Green and Smart Urban Development Guidelines" (China Development Bank Capital, October 2015), http://energyinnovation.org/wp-content/uploads/2015/12/12-Green-Guidelines.pdf.

3. Calthorpe & Associates, China Sustainable Transportation Center, and Glumac, *Emerald Cities: Planning for Smart and Green China* (Beijing: MOHURD Press, 2017).

4. Jason Margolis, "8 Million People. No Subway. Can This City Thrive without One?" *Public Radio International* (2015), https://www.pri.org/stories/2015-10-21/can-modern-megacity-bogot-get-without-subway.

5. Laura Bliss, "New York City Traffic Is Now 'Unsustainable,' Thanks to Ride-Hailing" *CityLab* (2017), accessed February 6, 2018, https://www.citylab.com/transportation/2017/12/how-to-fix-new-york-citys-unsustainable-traffic-woes/548798/.

6. "TOD Standard" (Institute for Transportation & Development Policy, 2014), https://www.itdp.org/wp-content/uploads/2014/03/The-TOD-Standard-2.1.pdf.

7. "History of the London Green Belt" (London Green Belt Council, 2018), accessed February 6, 2018, http://londongreenbeltcouncil.org.uk/history-of-the-london-green-belt/.

8. Floor area ratio (FAR) is the ratio of building floor space to the square footage of a parcel. Parcels commonly have a maximum allowed FAR, which serves as a limit on the density of development on that parcel.

9. "Designing the Bay Area's Second Transbay Rail Crossing" (SPUR, 2016), 17, http://www.spur.org/sites/default/files/publications-pdfs/SPUR-Designing-the-Bay-Area%27s-Second-Transbay-Rail-Crossing.pdf.

10. "Urban Growth Boundary" (Oregon Metro, 2014), https://www.oregonmetro.gov/urban-growth-boundary.

11. "Sustainable Communities" (California Air Resources Board), accessed December 19, 2017, https://www.arb.ca.gov/cc/sb375/sb375.htm.

12. "Time Is Money: The Economic Benefits of Transit Investment" (Chicago Metropolis 2020, 2007), 2, http://edrgroup.com/pdf/timeismoney.pdf.

13. Leah Harnack and Kim Kaiser, "Public–Private Partnerships" (Mass Transit, 2012), http://www.masstransitmag.com/article/10628016/public-private-partnerships.

14. Dan Malouff, "The US Has Only 5 True BRT Systems, and None Are 'Gold'" (Greater Greater Washington, 2013), https://ggwash.org/view/29962/the-us-has-only-5-true-brt-systems-and-none-are-gold.

15. Timothy Hurst, "Guangzhou's Remarkable Bus Rapid Transit System" *Reuters* (2011), https://www.reuters.com/article/idUS331644810020110405.

16. Claudia Gunter, "Guangzhou Opens Asia's Highest Capacity BRT System" (Institute

for Transportation and Development Policy, 2010), https://www.itdp.org/guangzhou-opens
-asias-highest-capacity-brt-system/.

17. Ibid.

18. "Planning for BRT-Oriented Development: Lessons and Prospects from Brazil and
Colombia" (Clean Air Institute, 2011), http://cleanairinstitute.org/download/folleto1-cai
.pdf.

19. "The Greatest Urban Experiment Right Now," *Copenhagenize* (blog), 2014, http:
//www.copenhagenize.com/2014/07/the-greatest-urban-experiment-right-now.html.

20. "Cycling in Copenhagen" (Wikipedia, 2017), https://en.wikipedia.org/w/index
.php?title=Cycling-in-Copenhagen&oldid=817743124.

21. "Strøget" (Wikipedia, 2017), https://en.wikipedia.org/w/index.php?title=Str%C3%
B8get&oldid=812886163.

22. "Congestion Charge (Official)" (Transport for London), accessed December 19, 2017,
https://www.tfl.gov.uk/modes/driving/congestion-charge.

23. "London Congestion Charge" (Wikipedia, 2017), https://en.wikipedia.org/w/index
.php?title=London-congestion-charge&oldid=813237659.

24. Sam Foss, *The Calf-Path*, 1895, Poetry, 1895, https://www.poets.org/poetsorg/poem
/calf-path.

Section III

1. World Resources Institute, "CAIT Climate Data Explorer" (2017), http://cait.wri.org.

2. Elmar Kriegler et al., "What Does the 2°C Target Imply for a Global Climate Agreement in 2020? The LIMITS Study on Durban Platform Scenarios" (Data downloaded from
the LIMITS database hosted at IIASA), https://tntcat.iiasa.ac.at/LIMITSDB/dsd?Action=
htmlpage&page=about.

3. Ibid. Demand measured as passenger-kilometers and freight-ton-kilometers.

4. Ibid.

Chapter 10

1. "International Energy Outlook 2016" (U.S. Energy Information Administration, 2016),
Table F1, https://www.eia.gov/outlooks/ieo/ieo-tables.php.

2. "World Urbanization Prospects" (United Nations Department of Economic and Social
Affairs, 2014), 20, https://esa.un.org/unpd/wup/publications/files/wup2014-highlights.Pdf.

3. *World Energy Outlook* (International Energy Agency, 2016) Annex A: World, http:
//www.worldenergyoutlook.org/publications/weo-2016/.

4. "World Urbanization Prospects," 1.

5. Michael Rosenberg, "Stable Whole Building Performance Method for Standard 90.1,"
ASHRAE Journal (2013): 33, http://eley.com/sites/default/files/pdfs/033-045-Rosenberg
-WEB.pdf.

6. Richard K. Lester and David M. Hard, *Unlocking Energy Innovation: How America Can
Build a Low-Cost, Low-Carbon Energy System* (Cambridge, MA: MIT Press, 2011), 96.

7. Jeffrey Rissman, "It Takes a Portfolio: A Broad Spectrum of Policies Can Best Halt Climate Change in Electricity Policy" *Electricity Policy* (2016), http://energyinnovation.org/wp
-content/uploads/2018/01/2016-08-18-Broad-Spectrum-Published-Article.pdf.

8. A discount rate reflects the time value of money. When consumers have high discount
rates, it means they value money more in the near term than in the long term. Conversely, a
low discount rate means consumers are indifferent to receiving money now or in later years.
In general, consumers tend to have high discount rates compared with other investment

mechanisms, meaning they tend to overvalue money in the short run, even it means they will earn less overall.

9. Rissman, "It Takes a Portfolio."

10. Amory Lovins, "Energy-Efficient Buildings: Institutional Barriers and Opportunities" (E Source, 1992), http://energyinnovation.org/wp-content/uploads/2014/12/Energy-Efficient -Buildings-Institutional-Barriers-and-Opportunities.pdf.

11. "Average Household Refrigerator Energy Use, Volume, and Price over Time" (Appliance Standards Awareness Project, 2016), http://www.appliance-standards.org/sites/default /files/refrigerator-graph-Nov-2016.pdf.

12. Marianne DiMascio, "How Your Refrigerator Has Kept Its Cool over 40 Years of Efficiency Improvements" (Appliance Standards Awareness Project, 2014), https://appliance -standards.org/blog/how-your-refrigerator-has-kept-its-cool-over-40-years-efficiency -improvements.

13. Jeffrey Rissman and Maxine Savitz, "Unleashing Private-Sector Energy R&D: Insights from Interviews with 17 R&D Leaders" (Energy Innovation Council, 2013), 32, http:// energyinnovation.org/wp-content/uploads/2014/06/unleashing-private-rd-jan2013.pdf.

14. Except for cases when eligible buyers fail to submit necessary claim forms or proof of purchase in order to receive the rebates to which they are entitled.

15. "Long-Standing Problems with DOE's Program for Setting Efficiency Standards Continue to Result in Forgone Energy Savings" (U.S. Government Accountability Office, 2007), http://www.gao.gov/new.items/d0742.pdf.

16. Nina Zheng et al., "Local Enforcement of Appliance Efficiency Standards and Labeling Program in China: Progress and Challenges" (Lawrence Berkeley National Laboratory, 2012), https://china.lbl.gov/sites/all/files/lbl-5743e-local-enforcement-aceeejune-2012.pdf.

17. Shui Bin and Steven Nadel, "How Does China Achieve a 95% Compliance Rate for Building Energy Codes? A Discussion about China's Inspection System and Compliance Rates" (American Council for an Energy-Efficient Economy, 2012), http://aceee.org/files /proceedings/2012/data/papers/0193-0002B61.pdf.

18. For example, see Shanti Pless et al., "How-To Guide for Energy-Performance-Based Procurement: An Integrated Approach for Whole Building High Performance Specifications in Commercial Buildings" (National Renewable Energy Laboratory, n.d.), https://www1.eere .energy.gov/buildings/publications/pdfs/rsf/performance-based-how-to-guide.pdf.

19. For an idea of the size and relative success of ESCO approaches in different countries, see "Energy Efficiency Market Report 2016" (International Energy Agency, 2015), 110–17, https://www.iea.org/eemr16/files/medium-term-energy-efficiency-2016-WEB.PDF.

20. "Property-Assessed Clean Energy (PACE) Financing of Renewables and Efficiency" (National Renewable Energy Laboratory, 2010), https://www.nrel.gov/docs/fy10osti/47097 .pdf.

21. "Inclusive Financing for Efficiency Upgrades" (Clean Energy Works, n.d.), https: //drive.google.com/file/d/0BzYyDNPW3cwwOFBzc3NyTTF2MEE/view; "On-Bill Energy Efficiency" (American Council for an Energy-Efficient Economy, 2012), https://aceee.org /sector/state-policy/toolkit/on-bill-financing.

22. OECD, "Green Investment Banks: Scaling Up Private Investment in Low-Carbon, Climate-Resilient Infrastructure" (OECD iLibrary, 2016), http://dx.doi.org/10.1787 /9789264245129-en.

23. "Title 24 Compliance" (Energy Performance Services), accessed December 19, 2017, http://www.title24express.com/what-is-title-24/title-24-compliance/.

24. "Background on the 2013 Building Energy Efficiency Standards" (California Energy

Commission), accessed December 19, 2017, http://www.energy.ca.gov/title24/2013stan
dards/background.html.

25. "2016 Building Energy Efficiency Standards for Residential and Nonresidential Buildings" (California Energy Commission, 2016), 30, http://www.energy.ca.gov/2015publica
tions/CEC-400-2015-037/CEC-400-2015-037-CMF.pdf.

26. "Building Energy Efficiency Standards Enforcement" (California Energy Commission), accessed December 19, 2017, http://www.energy.ca.gov/title24/enforcement/.

27. "2019 Building Energy Efficiency Standards" (California Energy Commission), accessed December 19, 2017, http://www.energy.ca.gov/title24/2019standards/index.html.

28. "Title 24, Part 6 Stakeholders" (California Energy Codes and Standards), accessed
December 19, 2017, http://title24stakeholders.com/.

29. "Public Participation in the Energy Efficiency Standards Update" (California Energy
Commission), accessed December 19, 2017, http://www.energy.ca.gov/title24/participation
.html.

30. "California's Energy Efficiency Standards Have Saved Billions" (California Energy
Commission), accessed December 19, 2017, http://www.energy.ca.gov/efficiency/savings
.html.

31. "Gross State Product" (California State Department of Finance), accessed December
19, 2017, http://www.dof.ca.gov/Forecasting/Economics/Indicators/Gross-State-Product/.

32. "California's 2016 Residential Building Energy Efficiency Standards" (California
Energy Commission, 2016), http://www.energy.ca.gov/title24/2016standards/rulemaking
/documents/2016-Building-Energy-Efficiency-Standards-infographic.pdf.

33. Ibid., 1.

34. Michael McNeil and Ana Maria Carreño, "Impacts Evaluation of Appliance Energy
Efficiency Standards in Mexico since 2000 Technical Report" (Super-Efficient Equipment
and Appliance Deployment Initiative, 2015), 4, http://www.superefficient.org/~/media/Files
/PublicationLibrary/2015/Impacts%20Evaluation%20of%20Appliance%20Energy%20
Efficiency%20Standards%20in%20Mexico%20since%202000.ashx.

35. Ibid., 4.

36. Ibid.

37. Ibid.

38. Ibid., 17–18.

39. Ibid., 4.

40. Ibid., 19.

41. Ibid., 21.

Section IV

1. World Resources Institute, "CAIT Climate Data Explorer" (2017), http://cait.wri.org.

2. Elmar Kriegler et al., "What Does the 2°C Target Imply for a Global Climate Agreement in 2020? The LIMITS Study on Durban Platform Scenarios" (Data downloaded from
the LIMITS database hosted at IIASA), https://tntcat.iiasa.ac.at/LIMITSDB/dsd?Action=
htmlpage&page=about.

3. Ibid.

4. Ibid.

5. World Resources Institute, "CAIT Climate Data Explorer," 2017.

6. Because of the effects of rounding, these figures together amount to 27 percent of
necessary total emission reductions (as seen in Figure S-4).

Chapter 11

1. The other sectors considered are electricity generation, transportation, and buildings (residential plus commercial).

2. "International Energy Outlook 2016" (U.S. Energy Information Administration, 2016), Table F1, https://www.eia.gov/outlooks/ieo/ieo-tables.cfm.

3. National Bureau of Statistics of China, "China Statistical Yearbook 2016," Table 9-9, http://www.stats.gov.cn/tjsj/ndsj/2016/html/0909EN.jpg.

4. "Myths and Facts about Industrial Opt-Out Provisions" (American Council for an Energy-Efficient Economy, 2016), http://aceee.org/sites/default/files/ieep-myths-facts.pdf.

5. "Industrial Efficiency Programs Can Achieve Large Energy Savings at Low Cost" (American Council for an Energy-Efficient Economy, 2016), http://aceee.org/sites/default /files/low-cost-ieep.pdf.

6. "Waste Heat Recovery: Technology and Opportunities in U.S. Industry" (U.S. Department of Energy, 2008), v, https://www1.eere.energy.gov/manufacturing/intensiveprocesses /pdfs/waste-heat-recovery.pdf.

7. Ibid., 17.

8. Ibid., 17.25.

9. Christina Galitsky and Ernst Worrell, "Energy Efficiency Improvement and Cost Saving Opportunities for the Vehicle Assembly Industry" (Lawrence Berkeley National Laboratory, 2008), https://www.energystar.gov/ia/business/industry/LBNL-50939.pdf.

10. "Energy-Efficiency Policy Opportunities for Electric Motor-Driven Systems" (International Energy Agency, 2011), https://www.iea.org/publications/freepublications/publica tion/energy-efficiency-policy-opportunities-for-electric-motor-driven-systems.html.

11. Galitsky and Worrell, "Energy Efficiency Improvement and Cost Saving Opportunities for the Vehicle Assembly Industry," 21.

12. "What Is a Variable Speed Drive?" (ABB, 2017), http://www.abb.com/cawp/db0003 db002698/a5bd0fc25708f141c12571f10040fd37.aspx.

13. Galitsky and Worrell, "Energy Efficiency Improvement and Cost Saving Opportunities for the Vehicle Assembly Industry," 21.

14. Ibid., 22–27.

15. Ibid., 30–32.

16. Christopher Null and Brian Caulfield, "Fade to Black: The 1980s Vision of 'Lights-Out' Manufacturing, Where Robots Do All the Work, Is a Dream No More," *CNN Money* (2003), http://money.cnn.com/magazines/business2/business2-archive/2003/06/01/343371 /index.htm.

17. Galitsky and Worrell, "Energy Efficiency Improvement and Cost Saving Opportunities for the Vehicle Assembly Industry," 21.

18. "SIMATIC Energy Manager PRO" (Siemens, 2017), https://www.siemens.com/global /en/home/products/automation/industry-software/automation-software/energymanage ment/simatic-energy-manager-pro.html.

19. Jenny Herzfeld, "The Value of Energy Management Systems and ISO 50001" (Clean Energy Ministerial, 2015), http://www.cleanenergyministerial.org/Our-Work/Initiatives /Appliances/Videos/the-value-of-energy-management-systems-and-iso-50001-42958.

20. Ibid.

21. "Circular Economy System Diagram" (Ellen MacArthur Foundation, 2017), https: //www.ellenmacarthurfoundation.org/circular-economy/interactive-diagram.

22. "Standards and Test Procedures" (U.S. Department of Energy, 2017), https://energy .gov/eere/buildings/standards-and-test-procedures.

23. United States, "Energy Conservation Program: Energy Conservation Standards for Commercial and Industrial Electric Motors; Final Rule," 10 CFR Part 431 § (2014), https://www.regulations.gov/document?D=EERE-2010-BT-STD-0027-0117.

24. "Industrial Efficiency Programs Can Achieve Large Energy Savings at Low Cost" (American Council for an Energy-Efficient Economy, 2016), http://aceee.org/sites/default/files/low-cost-ieep.pdf.

25. "Development and Implementation of Best Practices in Indian Foundry Industry" (Institute for Industrial Productivity, 2012), http://www.iipnetwork.org/development-and-implementation-best-practices-indian-foundry-industry.

26. "Energy Efficiency Revolving Fund (EERF), Thailand" (Institute for Industrial Productivity, 2012), http://www.iipnetwork.org/IIP-FinanceFactsheet-3-EERF.pdf.

27. "Growing Clean Energy Markets with Green Bank Financing" (Coalition for Green Capital, 2015), 2, http://coalitionforgreencapital.com/wp-content/uploads/2015/08/CGC-Green-Bank-White-Paper.pdf.

28. "Understanding Green Bonds" (The World Bank, 2009), http://treasury.worldbank.org/cmd/htm/Chapter-2-Understanding-Green-Bonds.html.

29. "China's GHG Emissions Reduction Policies" (Institute for Industrial Productivity, 2013), 1, http://www.iipnetwork.org/IIPFactSheet-China.pdf.

30. Ibid.

31. The World Bank, *A Cascade Decision-Making Approach: Infrastructure Finance: Guiding Principles for the World Bank Group* (Washington, DC: The World Bank, 2017), 3.

32. Database, "Energy and Carbon Intensity Targets of the 12th Five Year Plan" (Institute for Industrial Productivity, 2011), http://iepd.iipnetwork.org/policy/energy-and-carbon-intensity-targets-12th-five-year-plan.

33. "Insights into Industrial Energy Efficiency Policy Packages: Sharing Best Practices from Six Countries" (Institute for Industrial Productivity, 2012), 11, http://www.iipnetwork.org/InsightsIEE-IIP.pdf.

34. Ibid., 16.

35. "Energy Efficiency Benchmarking Covenant: Netherlands" (International Energy Agency, 2013), https://www.iea.org/policiesandmeasures/pams/netherlands/name-23862-en.php.

36. "Dutch Energy Efficiency Benchmarking Covenant: Results and Energy Tax Exemptions" (CE Delft, 2010), http://www.cedelft.eu/publicatie/dutch-energy-efficiency-benchmarking-covenant%3A-results-and-energy-tax-exemptions/1072.

37. "Reinventing Fire: Industry Executive Summary" (Rocky Mountain Institute, 2011), https://www.rmi.org/insights/reinventing-fire/reinventing-fire-industry/.

38. "EPA's Voluntary Methane Programs for the Oil and Natural Gas Industry" (U.S. EPA, 2016), https://www.epa.gov/natural-gas-star-program.

39. Herzfeld, "The Value of Energy Management Systems and ISO 50001."

40. Tamar Jacoby, "Why Germany Is So Much Better at Training Its Workers," *The Atlantic* (2014), https://www.theatlantic.com/business/archive/2014/10/why-germany-is-so-much-better-at-training-its-workers/381550/; and "Vocational Training in Germany—How Does It Work?" (Make It in Germany, 2017), http://www.make-it-in-germany.com/en/for-qualified-professionals/training-learning/training/vocational-training-in-germany-how-does-it-work.

41. Cai Yun, "China Stages the World's Largest Energy Saving Project" (Energy Foundation China, 2014), http://www.efchina.org/About-Us-en/Case-Studies-en/case-2014112606-en.

42. "Top-10,000 Energy-Consuming Enterprises Program" (Institute for Industrial

Productivity), 2011, http://iepd.iipnetwork.org/policy/top-10000-energy-consuming-enterprises-program.

43. Yun, "China Stages the World's Largest Energy Saving Project."

44. "Annual Data" (National Bureau of Statistics of China, 2012, 2016), http://www.stats.gov.cn/english/statisticaldata/AnnualData/. Energy chapter, Table 9 (for energy use); Table 9-16 (from 2016 yearbook) for energy intensity by GDP.

45. Gu Yang, "'Twelve Five' 10,000 Enterprises Exceeded the Energy-Saving Goals" (China Climate Change Info-Net, 2016), http://www.ccchina.gov.cn/Detail.aspx?newsId=58433&TId=57%22%20title=%22.

46. "United States Superior Energy Performance Program" (Institute for Industrial Productivity, 2012), http://www.iipnetwork.org/IIP-USA-SEP-factsheet.pdf.

47. Ibid.

48. "Superior Energy Performance" (U.S. Department of Energy), accessed December 20, 2017, https://www.energy.gov/eere/amo/superior-energy-performance.

49. "50001 Ready Program" (U.S. Department of Energy), accessed December 20, 2017, https://energy.gov/eere/amo/50001-ready-program.

50. "Insights into Industrial Energy Efficiency Policy Packages."

51. United States, "Standards of Performance for New Stationary Sources," 40 CFR Part 60 § (2011), https://www.gpo.gov/fdsys/pkg/CFR-2011-title40-vol6/xml/CFR-2011-title40-vol6-part60.xml.

52. "About Us" (Bulgarian Energy Efficiency and Renewable Sources Fund, 2017), http://www.bgeef.com/display.aspx?page=about.

53. "Bulgarian Energy Efficiency and Renewable Sources Fund—EERSF" (CITYnvest, n.d.), 4, http://citynvest.eu/sites/default/files/library-documents/Model%2019-Energy%20Efficiency%20and%20Renewable%20Sources%20Fund%20-EERSF-final.pdf.

54. Ibid., 4.

55. Ibid., 4.

56. Ibid., 4.

57. Ibid., 4.9.

Chapter 12

1. "California Targets Cow Gas, Belching and Manure as Part of Global Warming Fight," *Los Angeles Times* (2016), http://www.latimes.com/local/lanow/la-me-cow-gas-20161129-story.html.

2. "Enteric Fermentation—Greenhouse Gases," in *AP-42: Compilation of Air Emission Factors* (Washington, DC: U.S. Environmental Protection Agency, 2009), https://www3.epa.gov/ttnchie1/ap42/ch14/final/c14s04.pdf.

3. Keith Wagstaff, "Could Better Tech Prevent the Next Big Methane Leak?" *NBC News* (2016), https://www.nbcnews.com/tech/innovation/could-better-tech-prevent-next-big-methane-leak-n487566.

4. Phil McKenna, "Why Natural Gas May Be as Bad as Coal" *PBS Nova* (2015), http://www.pbs.org/wgbh/nova/next/earth/methane-regulations/.

5. "Green Completions" (IPIECA, 2014), http://www.ipieca.org/resources/energy-efficiency-solutions/units-and-plants-practices/green-completions/.

6. Robert Fares, "Methane Leakage from Natural Gas Production Could Be Higher Than Previously Estimated," *Scientific American* (blog), 2015, https://blogs.scientificamerican.com/plugged-in/methane-leakage-from-natural-gas-supply-chain-could-be-higher-than-previously-estimated/.

7. "Cement Industry Energy and CO_2 Performance: Getting the Numbers Right" (World Business Council for Sustainable Development, 2009), 22, http://www.wbcsdcement.org/pdf/CSI GNR Report final 18 6 09.pdf.

8. Michael J. Gibbs, Peter Soyka, and David Conneely, "CO_2 Emissions from Cement Production" (Intergovernmental Panel on Climate Change, 2001), http://www.ipcc-nggip.iges.or.jp/public/gp/bgp/3-1-Cement-Production.pdf.

9. "Clinker Substitution" (European Cement Association, 2013), http://lowcarboneconomy.cembureau.eu/index.php?page=clinker-substitution.

10. "China's Housing Sector Is Crumbling—Literally" (China Economic Review, 2014), http://www.chinaeconomicreview.com/china-housing-shoddy-building-quality-energy-incentives-GDP.

11. Sabine Zikeli et al., "Effects of Reduced Tillage on Crop Yield, Plant Available Nutrients and Soil Organic Matter in a 12-Year Long-Term Trial under Organic Management," *Sustainability* (2013), https://store.extension.iastate.edu/Product/Impact-of-Tillage-Crop-Rotation-Systems-on-Soil-Carbon-Sequestration-PDF; and C.S. Ofori, "The Challenge of Tillage Development in African Agriculture," in *Soil Tillage in Africa: Needs and Challenges* (Rome: Food and Agriculture Organization of the United Nations, 1993), Chapter 7, http://www.fao.org/docrep/t1696e/t1696e08.htm.

12. Daniela R. Carrijo, Mark E. Lundy, and Bruze A. Linquist, "Rice Yields and Water Use under Alternate Wetting and Drying Irrigation: A Meta-Analysis," *Field Crops Research* (2017), https://www.sciencedirect.com/science/article/pii/S0378429016307791.

13. "Basic Information about Landfill Gas" (U.S. EPA, 2016), https://www.epa.gov/lmop/basic-information-about-landfill-gas.

14. Ibid.

15. "Reducing the Impact of Wasted Food by Feeding the Soil and Composting" (U.S. EPA, 2015), https://www.epa.gov/sustainable-management-food/reducing-impact-wasted-food-feeding-soil-and-composting.

16. "Compost vs. Landfill" (Resource Recycling Systems, 2017), https://recycle.com/organics-compost-vs-landfill/.

17. "Refrigerants Environmental Data" (Linde Gases AG, n.d.), http://www.linde-gas.com/internet.global.lindegas.global/en/images/Refrigerants%20environmental%20GWPs17-111483.pdf.

18. "Frequent Questions about Coal Mine Methane" (U.S. EPA, 2015), https://www.epa.gov/cmop/frequent-questions.

19. Ibid.

20. Ibid.

21. Ibid.

22. "Municipal Wastewater Methane: Reducing Emissions, Advancing Recovery and Use Opportunities" (Global Methane Initiative, 2013), https://www.globalmethane.org/documents/ww_fs_eng.pdf.

23. Ibid.

24. Ibid.

25. "Enteric Fermentation—Greenhouse Gases."

26. Jeanette Fitzsimons, "Can We Make Steel without Coal?" (Coal Action Network Aotearoa, 2013), http://coalaction.org.nz/carbon-emissions/can-we-make-steel-without-coal.

27. Ibid.

28. Meredith MacLeod, "U.S. Steel: Natural Gas Process Will Soon Replace Coke," *The*

Hamilton Spectator (2013), sec. News-Business, https://www.thespec.com/news-story /4190319-u-s-steel-natural-gas-process-will-soon-replace-coke/; and Bowdeya Tweh, "U.S. Steel to Reduce Coke, Use Natural Gas," *Northwest Indiana Times* (2011), http://www .nwitimes.com/niche/inbusiness/newsletter-featured-articles/u-s-steel-to-reduce-coke-use -natural-gas/article-b8b9f34d-a0fe-5671-a653-b1cd4a2cdb3a.html.

29. "Buy Clean California Act," AB-262, 2017, https://leginfo.legislature.ca.gov/faces /billNavClient.xhtml?bill-id=201720180AB262.

30. "The Montreal Protocol on Substances That Deplete the Ozone Layer" (U.S. Department of State, 2016), http://www.state.gov/e/oes/eqt/chemicalpollution/83007.htm.

31. Stephen Leahy, "Without the Ozone Treaty You'd Get Sunburned in 5 Minutes" *National Geographic News* (2017), https://news.nationalgeographic.com/2017/09/montreal -protocol-ozone-treaty-30-climate-change-hcfs-hfcs/.

32. "International Day for the Preservation of the Ozone Layer, 16 September" (United Nations, 2016), https://www.un.org/en/events/ozoneday/background.shtml.

33. "Montreal Protocol" (Wikipedia, 2017), https://en.wikipedia.org/w/index .php?title=Montreal-Protocol&oldid=808281366.

34. "International Treaties and Cooperation" (U.S. EPA, 2015), https://www.epa.gov /ozone-layer-protection/international-treaties-and-cooperation.

35. "International Day for the Preservation of the Ozone Layer."

36. "DuPont: A Case Study in the 3D Corporate Strategy" (Greenpeace, 1997), https: //web.archive.org/web/20120406093303/http://archive.greenpeace.org/ozone/greenfreeze /moral97/6dupont.html; Jeffrey Masters, "The Skeptics vs. the Ozone Hole" (Weather Underground), accessed December 22, 2017, https://www.wunderground.com/resources/cli mate/ozone-skeptics.asp; and Jack Doyle, "DuPont's Disgraceful Deeds: The Environmental Record of E.I. DuPont de Nemour," *The Multinational Monitor* 12, no. 10, accessed December 22, 2017, http://www.multinationalmonitor.org/hyper/issues/1991/10/doyle.html.

37. "Discovering the Ozone Hole: Q&A with Pawan Bhartia" (NASA, 2012), https://www .nasa.gov/topics/earth/features/bhartia-qa.html.

38. Leahy, "Without the Ozone Treaty You'd Get Sunburned in 5 Minutes."

39. David Doniger, "Trump Budget Attacks Montreal Protocol, Reagan's Crown Jewel" (NRDC, 2017), https://www.nrdc.org/experts/david-doniger/trump-budget-attacks-mon treal-protocol-reagans-crown-jewel.

40. "SNAP Regulations" (U.S. EPA, 2014), https://www.epa.gov/snap/snap-regulations.

41. United States, "Protection of Stratospheric Ozone: Change of Listing Status for Certain Substitutes under the Significant New Alternatives Policy Program," 40 CFR Part 82 § (2015), 42876–77, https://www.gpo.gov/fdsys/pkg/FR-2015-07-20/pdf/2015-17066.pdf.

42. "Landmark Revision of Japanese Fluorocarbon Regulations: Make It Count" (Environmental Investigation Agency, 2013), https://eia-global.org/blog-posts/landmark-revision -of-japanese-fluorocarbon-regulations-make-it-count-1.

43. Atsuhiro Meno, "Laws and Regulation for Fluorocarbons in Japan" (Ministry of Economy, Trade and Industry, Japan, 2015), https://www.jraia.or.jp/english/icr/ICR2015 -METI.pdf.

44. Ibid.

45. "Dairy Digester Research and Development Program (DDRDP)" (California State Department of Food and Agriculture), accessed December 23, 2017, https://www.cdfa.ca .gov/oefi/ddrdp/.

46. Ibid.

47. "2017 Dairy Digester Research and Development Program: Projects Selected for

Award of Funds" (California State Department of Food and Agriculture, 2017), https://www
.cdfa.ca.gov/oefi/ddrdp/docs/2016-Fact-Sheet.pdf.

48. "2017 Dairy Digester Research and Development Program."

49. "National Domestic Biogas Programme—Zimbabwe," (SNV, n.d.), accessed December 23, 2017, http://www.snv.org/project/national-domestic-biogas-programme-zimbabwe.

50. Ibid.

51. Felix Ter Heegde and Kai Sonder, "Biogas for a Better Life: An African Initiative" (SNV, 2007), http://www.snv.org/public/cms/sites/default/files/explore/download/20070520
-biogas-potential-and-need-in-africa.pdf.

52. Ibid.

53. "National Domestic Biogas Programme—Zimbabwe."

54. Ibid.

55. "Forging Partnerships for Renewable Energy in Zimbabwe" (United Nations Development Programme, 2014), http://www.zw.undp.org/content/zimbabwe/en/home/presscenter
/articles/2014/11/10/forging-partnerships-for-renewable-energy-in-zimbabwe.html.

56. "Biogas Digesters Promote Clean Energy" (Hivos, 2015), https://www.hivos.org/news
/biogas-digesters-promote-clean-energy.

57. "EPA's Actions to Reduce Methane Emissions from the Oil and Natural Gas Industry: Final Rules and Draft Information Collection Request" (U.S. EPA, 2016), https://www.epa
.gov/sites/production/files/2016-09/documents/nsps-overview-fs.pdf.

58. An organic vapor analyzer detects volatile organic compounds (VOCs), not methane, but VOC leaks are associated with methane leaks.

59. "EPA's Actions to Reduce Methane Emissions from the Oil and Natural Gas Industry."

60. "Green Completions."

61. Ibid.

62. "EPA's Actions to Reduce Methane Emissions from the Oil and Natural Gas Industry."

Chapter 13

1. Leakage occurs when emissions or economic activities shift outside an area because of the policy. Leakage to neighboring places is the greatest likelihood, because these are often the greatest trading partners, although in the era of globalization leakage is not constrained to neighboring countries or states.

2. This line of analysis goes back to Martin L. Weitzman, "Prices vs. Quantities," *The Review of Economic Studies* 41, no. 4 (1974): 477–91, https://doi.org/10.2307/2296698.

3. Hal Berton, "Washington State Alliance to Push a Reworked Carbon-Tax Proposal," *The Seattle Times* (November 12, 2016), https://www.seattletimes.com/seattle-news/envi
ronment/washington-state-alliance-to-push-a-reworked-carbon-tax-initiative/.

4. Marc Hafstead, Gilbert E. Metcalf, and Roberton C. Williams III, "Adding Quantity Certainty to a Carbon Tax: The Role of a Tax Adjustment Mechanism for Policy Pre-Commitment" (Resources for the Future, 2016), http://www.rff.org/research/publications/add
ing-quantity-certainty-carbon-tax-role-tax-adjustment-mechanism-policy-pre; and Brian Murray, William A. Pizer, and Christina Reichert, "Increasing Emissions Certainty under a Carbon Tax" (Duke Nicholas Institute for Environmental Policy Solutions, 2016), https:
//nicholasinstitute.duke.edu/sites/default/files/publications/ni-pb-16-03.pdf.

5. For more reading, see "Tackling Carbon Leakage" (UK Carbon Trust; Climate Strategies, 2010), https://www.carbontrust.com/media/84908/ctc767-tackling-carbon-leakage.pdf.

6. Quote from "Allocating Emissions Allowances under a California Cap-and-Trade Program" (California Economic and Allocation Advisory Committee, 2010), 13, http://www

.climatechange.ca.gov/eaac/documents/eaac-reports/2010-03-22-EAAC-Allocation-Report
-Final.pdf.

7. Dallas Burtraw, Richard Sweeney, and Margaret Walls, "The Incidence of U.S. Climate
Policy: Alternative Uses of Revenues from a Cap-and-Trade Auction" (Washington, DC: Resources for the Future, June 2009), http://www.rff.org/documents/RFF-DP-09-17-REV.pdf.

8. For some excellent discussion, see Meredith Fowlie, "The Promise and Perils of Linking Carbon Markets" (Energy Institute Blog, 2016), https://energyathaas.wordpress.com
/2016/07/25/the-promise-and-perils-of-linking-carbon-markets/.

9. Esther Duflo et al., "Truth-Telling by Third-Party Auditors and the Response of Polluting Firms: Experimental Evidence from India," *The Quarterly Journal of Economics* 128, no. 4
(November 1, 2013): 1499–1545, https://doi.org/10.1093/qje/qjt024.

10. M. Jeff Hamond et al., *Tax Waste, Not Work: How Changing What We Tax Can Lead
to a Stronger Economy and a Cleaner Environment* (San Francisco, CA: Redefining Progress,
1997).

11. "State and Trends of Carbon Pricing 2017" (World Bank; Ecofys; Vivid Economics,
2017), https://openknowledge.worldbank.org/handle/10986/28510.

12. Some critiques argue that this figure is much too low. The following study estimates
the true value as $220 per ton: Frances C. Moore and Delavane B. Diaz, "Temperature Im-
pacts on Economic Growth Warrant Stringent Mitigation Policy," *Nature Climate Change* 5
(January 12, 2015): 127.

13. "The Investment of RGGI Proceeds through 2014" (Regional Greenhouse Gas Initiative, 2016), https://www.rggi.org/docs/ProceedsReport/RGGI-Proceeds-Report-2014.pdf.

14. Paul Hibbard et al., "The Economic Impacts of the Regional Greenhouse Gas Initiative on Nine Northeast and Mid-Atlantic States" (Analysis Group, 2015), http://www.analy
sisgroup.com/uploadedfiles/content/insights/publishing/analysis-group-rggi-report-july
-2015.pdf.

15. "Analysis of the Public Health Impacts of the Regional Greenhouse Gas Initiative,
2009–2014" (Abt Associates, 2017), http://www.abtassociates.com/AbtAssociates/files/7e
/7e38e795-aba2-4756-ab72-ba7ae7f53f16.pdf.

16. Michael Grubb and Federico Ferrario, "False Confidences: Forecasting Errors and
Emission Caps in CO_2 Trading Systems," *Climate Policy* 6, no. 4 (January 1, 2006): 495–501,
https://doi.org/10.1080/14693062.2006.9685615.

17. "RGGI States Announce Proposed Program Changes: Additional 30% Emissions Cap
Decline by 2030" (RGGI Inc., 2017), http://rggi.org/docs/ProgramReview/2017/08-23-17
/Announcement-Proposed-Program-Changes.pdf.

18. Ibid.

19. Dallas Burtraw, "Evaluating Experience with the Cost-Containment Reserve & Ideas
for the Future" (RGGI Inc., 2016), https://www.rggi.org/docs/ProgramReview/2016/04-29
-16/Burtraw-on-RGGI-CCR-April-29th.pdf.

20. Chris Busch, "Oversupply Grows in the Western Climate Initiative Carbon Market"
(Energy Innovation, 2017), http://energyinnovation.org/wp-content/uploads/2017/12
/Oversupply-Grows-In-The-WCI-Carbon-Market.pdf.

Chapter 14

1. "A Business Plan for America's Energy Future" (American Energy Innovation Council,
2010), 10, http://bpcaeic.wpengine.com/wp-content/uploads/2012/04/AEIC-The-Business
-Plan-2010.pdf.

2. Ibid.

3. "Research and Development Expenditure (% of GDP)" (World Bank, n.d.), https://data
.worldbank.org/indicator/GB.XPD.RSDV.GD.ZS?year-high-desc=true.

4. Charlie Taylor, "Most R&D Activities in Ireland Carried Out by Multinationals," *Irish
Times* (2016), https://www.irishtimes.com/business/most-r-d-activities-in-ireland-carried
-out-by-multinationals-1.2727089; and Vivian E. Nathan, "Reasons to Do Business in Ireland—Research and Development (R&D) Tax Credits," *Roberts Nathan* (blog), 2015, http:
//www.robertsnathan.com/reasons-to-do-business-in-ireland-research-and-development
-rd-tax-credits/.

5. Stephen Lacey, "DOE: The Clean Energy Loan Program Is Already Making a Profit for
Taxpayers," *Greentech Media* (2014), https://www.greentechmedia.com/articles/read/the
-clean-energy-loan-program-is-already-making-money-for-taxpayers.

6. "Research & Experimentation Tax Credit" (Wikipedia, 2017), https://en.wikipedia.org
/w/index.php?title=Research-%26-Experimentation-Tax-Credit&oldid=794853537.

7. Tom Sanger, "R&D Tax Credit: New, Improved, and Permanent," *CFO* (blog), 2016,
http://ww2.cfo.com/tax/2016/03/rd-tax-credit-new-improved-permanent/.

8. "The Quadrennial Technology Review" (U.S. Department of Energy, 2015), https://
energy.gov/under-secretary-science-and-energy/quadrennial-technology-review.

9. "Stage-Gate Innovation Management Guidelines: Managing Risk through Structured
Project Decision-Making" (U.S. Department of Energy, 2007), https://www1.eere.energy
.gov/manufacturing/financial/pdfs/itp-stage-gate-overview.pdf.

10. "About Us: History" (Central Power Research Institute, 2012), http://www.cpri.in
/about-us/about-cpri/history.html.

11. "Technology Deployment Centers" (Sandia National Laboratories, n.d.), accessed
January 3, 2018, http://www.sandia.gov/research/facilities/technology-deployment-centers
/index.html.

12. Bradley P. Nelson, "Patent Trolls: Can You Sue Them for Suing or Threatening to Sue
You?" (American Bar Association, 2014), http://apps.americanbar.org/litigation/committees
/intellectual/articles/fall2014-0914-patent-trolls-can-you-sue-them.html.

13. Kenneth L. Bressler, Christopher K. Hu, and Thomas H. Belknap, Jr., "Patent Trolls:
How to Defend against Patent Trolls without Breaking the Bank" (Blank Rome LLP, 2015),
https://www.blankrome.com/index.cfm?contentID=37&itemID=3703.

14. The use of patents not to protect one's own IP but to support a countersuit as a deterrent to patent litigation is so widespread that Google has initiated a program, PatentShield,
that gives startups access to Google's vast collection of patents that they can use to countersue their opponents, in return for a stake in those companies. For more information, see
Frederic Lardinois, "Google's and Intertrust's New PatentShield Helps Startups Fight Patent
Litigation in Return for Equity" *TechCrunch* (2017), http://social.techcrunch.com/2017/04
/25/googles-and-intertrusts-new-patentshield-helps-startups-fight-patent-litigation-in-re-
turn-for-equity/.

15. One example is United for Patent Reform, whose members include Amazon, AT&T,
Facebook, Ford, GM, Google, Macy's, the National Association of Broadcasters, the National
Restaurant Association, the National Retail Federation, Starwood Hotels, Walmart, and
many others.

16. Jeffrey Rissman and Maxine Savitz, "Unleashing Private-Sector Energy R&D: Insights
from Interviews with 17 R&D Leaders" (Energy Innovation Council, 2013), 38. http://energy
innovation.org/wp-content/uploads/2014/06/unleashing-private-rd-jan2013.pdf.

17. Ibid., 33.

18. Ibid., 40.

19. Jennifer Hunt, "Analysis: Would the U.S. Benefit from a Merit-Based Immigration System?" *PBS NewsHour* (2017), https://www.pbs.org/newshour/economy/analysis-u-s -benefit-merit-based-immigration-system.

20. Note that these countries also admit immigrants via nonpoint systems, such as for family ties or humanitarian reasons.

21. "Bridging the Gap, Powered by Ideas" (Defense Advanced Research Projects Agency, 2005), http://www.dtic.mil/cgi-bin/GetTRDoc?Location=U2&doc=GetTRDoc .pdf&AD=ADA433949.

22. "About" (Advanced Research Projects Agency–Energy), accessed January 3, 2018, https://arpa-e.energy.gov/?q=arpa-e-site-page/about.

23. Ibid.

24. "ARPA-E Impact" (Advanced Research Projects Agency–Energy), accessed January 3, 2018, https://arpa-e.energy.gov/?q=site-page/arpa-e-impact.

25. "ARPA-E: The First Seven Years: A Sampling of Project Outcomes" (Advanced Research Projects Agency–Energy, 2016), https://arpa-e.energy.gov/sites/default/files/docu ments/files/Volume%201-ARPA-E-ImpactSheetCompilation-FINAL.pdf.

26. "Introduction" (Innovation Network Corporation of Japan, n.d.), accessed January 3, 2018, https://www.incj.co.jp/english/.

27. "Investment Activities" (Innovation Network Corporation of Japan, n.d.), https: //www.incj.co.jp/investment/index.html.

28. Ibid.

29. "Investment Case List (Energy)" (Innovation Network Corporation of Japan, n.d.), http://www.incj.co.jp/investment/deal-list2.html?cat1=04.

30. "GE Joins New Government-Led Initiative in Japan to Accelerate Technology Innovation" (General Electric, 2009), http://www.genewsroom.com/press-releases/ge -joins-new-government-led-initiative-in-japan-to-accelerate------technology-innovation -244876.

31. "Cooperation with External Organizations" (Innovation Network Corporation of Japan, n.d.), http://www.incj.co.jp/openinnovation/collaboration.html.

32. Charles W. Wessner, "How Does Germany Do It?" (American Society of Mechanical Engineers, 2013), https://www.asme.org/engineering-topics/articles/manufacturing-pro cessing/how-does-germany-do-it.

33. "Facts and Figures (February 2017)" (Fraunhofer-Gesellschaft, 2017), https://www .fraunhofer.de/en/about-fraunhofer/profile-structure/facts-and-figures.html.

34. "Finances" (Fraunhofer-Gesellschaft, 2017), https://www.fraunhofer.de/en/about -fraunhofer/profile-structure/facts-and-figures/finances.html.

35. "Facts and Figures (February 2017)."

36. "Groups" (Fraunhofer-Gesellschaft, 2017), https://www.fraunhofer.de/en/about -fraunhofer/profile-structure/structure-organization/fraunhofer-groups.html.

37. Wessner, "How Does Germany Do It?"

38. Ibid.

Chapter 15

1. "Global Carbon Budget 2017" (Global Carbon Project, 2017), http://www.global carbonproject.org/carbonbudget/17/files/GCP-CarbonBudget-2017.pdf.

2. Nicolette Fox, "David Ball Group Invents Cement-Free Concrete" *The Guardian* (2015), https://www.theguardian.com/sustainable-business/2015/apr/30/david-ball-group -invents-cement-free-concrete.

3. "Large-Scale CCS Facilities" (Global CCS Institute, n.d.), https://www.globalccsinstitute.com/projects/large-scale-ccs-projects.

4. "Allam Power Cycle" (Wikipedia, n.d.), https://en.wikipedia.org/wiki/Allam-power-cycle; and "Technology" (NetPower, n.d.), https://www.netpower.com/technology/.

5. Chelsea Harvey, "We're Placing Far Too Much Hope in Pulling Carbon Dioxide out of the Air, Scientists Warn," *Washington Post* (2016), https://www.washingtonpost.com/news/energy-environment/wp/2016/10/13/were-placing-far-too-much-hope-in-pulling-carbon-dioxide-out-of-the-air-scientists-warn/.

6. "In-Depth: Experts Assess the Feasibility of 'Negative Emissions,'" *Carbon Brief* (April 12, 2016), https://www.carbonbrief.org/in-depth-experts-assess-the-feasibility-of-negative-emissions.

7. "Direct Air Capture of CO_2 with Chemicals: A Technology Assessment of the APS Panel on Public Affairs" (American Physical Society, 2011), https://www.aps.org/policy/reports/assessments/upload/dac2011.pdf.

8. Daniel Cressey, "Rock's Power to Mop Up Carbon Revisited," *Nature* (2014), https://www.nature.com/news/rock-s-power-to-mop-up-carbon-revisited-1.14560.

9. "Limestone: What Is Limestone and How Is It Used?" (Geology.com, n.d.), https://geology.com/rocks/limestone.shtml.

10. Cressey, "Rock's Power to Mop Up Carbon Revisited."

11. Suzanne J.T. Hangx and Christopher J. Spiers, "Coastal Spreading of Olivine to Control Atmospheric CO_2 Concentrations: A Critical Analysis of Viability," *International Journal of Greenhouse Gas Control* 3, no. 6 (December 2009): 757–67, http://www.sciencedirect.com/science/article/pii/S1750583609000656?via%3Dihub.

12. "Iron Fertilization" (Wikipedia, n.d.), https://en.wikipedia.org/wiki/Iron-fertilization.

13. "The Importance of Phytoplankton" (NASA, n.d.), https://earthobservatory.nasa.gov/Features/Phytoplankton/page2.php.

14. Ibid.

15. Columbia University, "Seeding Iron in the Pacific May Not Pull Carbon from Air as Thought" (Phys.org, 2016), https://phys.org/news/2016-03-seeding-iron-pacific-carbon-air.html.

16. "Biofuels versus Gasoline: The Emissions Gap Is Widening" (Environmental and Energy Study Institute, 2016), http://www.eesi.org/articles/view/biofuels-versus-gasoline-the-emissions-gap-is-widening.

17. Eric Wesoff, "Hard Lessons from the Great Algae Biofuel Bubble" *Green Tech Media* (2017), https://www.greentechmedia.com/articles/read/lessons-from-the-great-algae-biofuel-bubble.

18. "Why Solar Fuels?" (Joint Center for Artificial Photosynthesis, n.d.), https://solarfuelshub.org/why-solar-fuels/.

19. "World's First Hydrogen-Powered Cruise Ship Scheduled" (Maritime Executive, 2017), https://maritime-executive.com/article/worlds-first-hydrogen-powered-cruise-ship-scheduled.

20. Marua Judkis, "Could Congress Put Horsemeat Back on the Menu in America?," *The Washington Post* (2017), https://www.washingtonpost.com/news/food/wp/2017/07/14/could-congress-put-horsemeat-back-on-the-menu-in-america/?utm-term=.e6a378a50382.

21. "Geoengineering" (The Keith Group at Harvard University, n.d.), https://keith.seas.harvard.edu/geoengineering.

22. Ibid.

23. Hashem Akbari et al., "The Long-Term Effect of Increasing the Albedo of Urban

Areas," *Environmental Research Letters* (2012), http://iopscience.iop.org/article/10.1088/1748-9326/7/2/024004/meta.

24. "WG2 Chapter 14: Adaptation" (IPCC, 2015), https://www.ipcc.ch/pdf/assessment-report/ar5/wg2/WGIIAR5-Chap14-FINAL.pdf.

25. Melita Sunjic, "Top UNHCR Official Warns about Displacement from Climate Change" (UNHCR, 2008), http://www.unhcr.org/493e9bd94.html.

26. Jessica Benko, "How a Warming Planet Drives Human Migration," *The New York Times* (2017), https://www.nytimes.com/2017/04/19/magazine/how-a-warming-planet-drives-human-migration.html.

27. Michael Werz and Laura Conley, "Climate Change, Migration, and Conflict: Tackling Complex Crisis Scenarios in the 21st Century," *American Progress* (2012), https://cdn.americanprogress.org/wp-content/uploads/issues/2012/01/pdf/climate-migration.pdf.

Conclusion

1. Justin Gillis and Hal Harvey, "Why a Big Utility Is Embracing Wind and Solar," *The New York Times* (2018), sec. Opinion, https://www.nytimes.com/2018/02/06/opinion/utility-embracing-wind-solar.html; and Silvio Marcacci, "India Coal Power Is About to Crash: 65% of Existing Coal Costs More Than New Wind and Solar," *Forbes*, accessed (2018), https://www.forbes.com/sites/energyinnovation/2018/01/30/india-coal-power-is-about-to-crash-65-of-existing-coal-costs-more-than-new-wind-and-solar/.

2. "Shenzhen Completes Switch to Fully Electric Bus Fleet. Electric Taxis Are Next." *CleanTechnica* (2018), https://cleantechnica.com/2018/01/01/shenzhen-completes-switch-fully-electric-bus-fleet-electric-taxis-next/.

Appendix I

1. The Energy Policy Simulator includes nonenergy policies, such as those affecting land use and industrial processes, as well as energy policies.

2. For more information, see "Energy Policy Simulator Documentation," January 10, 2018, https://us.energypolicy.solutions/docs/.

3. The Energy Policy Simulator calculates first-order costs and savings (which entities pay which other entities more or less money relative to the business-as-usual case). It does not include a full macroeconomic simulation of the economy, which would consider questions such as how government spends increased tax revenues (or what government cuts in response to reduced tax revenues), because these decisions are beyond the scope of the energy and nonenergy policies that the computer model seeks to evaluate.

4. Macroeconomic models may be particularly useful for creating a projected business-as-usual case, because their strength is in elucidating economic interactions, but they may have trouble representing certain policies, particularly those that save money by causing actions that aren't undertaken in the absence of policy, because of market failures, irrational behavior by economic actors, nonmarket barriers, and so on. Technology-based models may be very useful for understanding the maximum potential abatement that could be derived from different sectors or different activities, which is helpful when deciding which sectors or activities to target with policies. However, they may not provide insight into what policies would induce those technical changes. A system dynamics model is strong at estimating how policies would affect emissions, cash flows, and so on relative to the business-as-usual case, without needing to rely on many of the underlying assumptions used by macroeconomic models.

5. For more information on how each sector functions, see the model's online

documentation at "Energy Policy Simulator Documentation," https://us.energypolicy.solutions/docs/.

6. Agriculture is handled with other industries because of similarities in how its emissions can be handled within the model structure. For example, agricultural equipment burns fuel, just as fuel is burned by machinery in manufacturing industries. Similarly, just as other industries have process emissions unrelated to fuel combustion, agriculture has such emissions (e.g., methane from rice fields or ruminant animals). Aspects of agriculture related to land use change, such as cutting down a forest to create a plantation, are handled in the "Land Use" section of the model, not the "Industry and Agriculture" section.

7. Most commonly, scaled data originate from the United States. The types of data missing in many countries tend to be the same, whereas U.S. data availability is relatively good.

Appendix II

1. "About LIMITS" (Science for Global Insight, n.d.), https://tntcat.iiasa.ac.at/LIMITS PUBLICDB/dsd?Action=htmlpage&page=about.

2. "LIMITS Work Package 1–2°C Scenario Study Protocol" (Science for Global Insight, n.d.), https://tntcat.iiasa.ac.at/LIMITSPUBLICDB/static/download/LIMITS-overview -SOM-Study-Protocol-Final.pdf.

3. World Resources Institute, "CAIT Climate Data Explorer" (2017), http://cait.wri.org.

Index

Note: Page numbers followed by the letter f or t indicate figures or tables

About the Authors

Hal Harvey is the CEO of Energy Innovation: Policy and Technology LLC, a San Francisco-based energy policy firm. He was the founder of ClimateWorks Foundation and Energy Foundation, and served as Environment Program Director at the William and Flora Hewlett Foundation. He served on energy panels appointed by Presidents Bush (41) and Clinton, and currently serves as a board member for several financial, science, and philanthropic groups. In 2017, Hal received the Heinz Award for his long track-record fighting climate change. Hal has BS and MS degrees from Stanford University in Engineering, specializing in Energy Planning.

Robbie Orvis is Director of Energy Policy Design at Energy Innovation: Policy and Technology LLC. He has worked with numerous governments on climate and energy policy, including Canada, China, Indonesia, Mexico, Poland and the US, with a focus on power sector decarbonization and building energy efficiency. His work has been featured in *Forbes, The New York Times, The Washington Post,* and other publications. Robbie holds a Master of Environmental Management from Yale University and a BS from UC Berkeley.

Jeffrey Rissman is the Industry Program Director and Head of Modeling at Energy Innovation: Policy and Technology LLC. He has worked with governments and researchers on climate and energy policy in many countries, including China, Canada, India, Indonesia, Mexico, and the US, focusing on long-term modeling and industrial sector decarbonization. His work has been featured in *Forbes, Bloomberg, The New York Times,* and other publications. Jeffrey holds an MS in Environmental Sciences and Engineering and a Masters of City and Regional Planning from UNC Chapel Hill.